PRAISE FOR
LONG FOR THIS WORLD

"Insightful [and] engaging. . . . Weiner explores the fractured, fuzzy science and pseudoscience of immortality. . . . *Long for This World* is a great trip."
—*New York Times Book Review*

"An astute and elegantly presented discussion about why all living things age and die and how life would change were it never to end.... [A] fascinating, deeply thought-provoking book full of intricate science and complicated moral questions made easily accessible for...us ordinary mortals."
—*Philadelphia Inquirer*

"[A]n extraordinary ride.... The author is as adept at parsing the ancient epic of Gilgamesh...as he is at explaining the inner workings of mitochondria."
—Bloomberg.com

"A brilliant and improbably funny look inside the mind-bending science of immortality.... [Weiner's] ability to write simply and swiftly about complex...processes makes him an ideal guide through the burgeoning field of gerontology, or the science of longevity."
—*Village Voice*

"*Long for This World* is a brilliant exposé of the fascinating science that has emerged in the quest for everlasting life.... The genius of Weiner's book is in the storytelling."
—*New Scientist*

"[I]nteresting and arresting facts permeate the book.... The science here is difficult, but it is also where Jonathan Weiner is at his best, clearly and cogently explaining the cryptic biology behind our mortality." —*Commentary*

"Fascinating.... Pulitzer Prize–winning science writer Jonathan Weiner takes readers to the cutting edge of medical research that some experts say could make aging a 'curable disease.'" —*Parade*

"[G]ripping.... In this wonderfully crafted book, Weiner explores the history of humankind's yearning for longevity.... Weiner's lucid, brightly paced narrative brims with snapshots of scientists, stories of experiments and informed speculations on what the conquest of aging would mean for the human experience. Immensely readable and informative." —*Kirkus Reviews*

"An engrossing tour of cutting-edge research.... [H]e has a knack for translating science into evocative metaphor.... Weiner's erudite, elegant exposition of the underlying science is stimulating yet sobering."
—*Publishers Weekly* (starred review)

"In the hands of a gifted writer like Jonathan Weiner, man's quest for immortality becomes illuminating and inspiring.... It is a science book, but one written with verve and vitality... [A] book that celebrates mankind's imagination, inventiveness and inspiration."
—Bookpage

"Pulitzer Prize–winning Weiner delves into [Aubrey] de Grey's hyperbole about extending human life spans by hundreds of years.... Is such a goal even desirable or ethical? Weiner weaves his answers with his own musings and those from religious and intellectual traditions, and leaves readers with an indelible impression." —*Booklist*

"Jonathan Weiner has done it again. In *Long for This World,* one of our finest science journalists explores the shadowy sword hanging over us all, weaving together the latest research with time-tested cultural wisdom. Will we ever live forever? And would we even want to?"

—Jonah Lehrer

"In *Long for This World,* Jonathan Weiner brings his immense talents—his masterful prose, his deep reporting, and his ability to see connections across the human experience—to one of science's most intriguing frontiers: the science of aging."

—Carl Zimmer

"Taxes may be inevitable, but death? Maybe not so much, suggests Jonathan Weiner, one of our finest science writers, in this searching and surprisingly witty look at the scientific odds against tomorrow."

—Timothy Ferris

"I admire all of Jonathan Weiner's books, but this one especially because of its intellectual depth and clarity, its sense of personal involvement, and its tone and wit. The chapter on the evolution of aging is particularly brilliant! I couldn't put the book down."

—Oliver Sacks

ALSO BY JONATHAN WEINER

His Brother's Keeper

Time, Love, Memory

The Beak of the Finch

The Next One Hundred Years

Planet Earth

THE STRANGE SCIENCE OF IMMORTALITY

JONATHAN WEINER

An Imprint of HarperCollinsPublishers

For my father

HarperCollins books may be purchased for educational, business, or sales promotional use. For information please write: Special Markets Department, HarperCollins Publishers, 10 East 53rd Street, New York, NY 10022.

A hardcover edition of this book was published in 2010 by Ecco, an imprint of HarperCollins Publishers.

FIRST ECCO PAPERBACK EDITION PUBLISHED 2011.

Designed by Suet Yee Chong

The Library of Congress has cataloged the hardcover edition as follows:

Weiner, Jonathan.
Long for this world : the strange science of immortality / Jonathan Weiner.—1st ed.
p. cm.
Includes bibliographical references (p. 285-297) and index.
ISBN 978-0-06-076536-1
1. Immortalism. 2. Longevity. 3. Immortality (Philosophy) 4. Science—Miscellanea. I. Title.

[call number n/a]
612.68 22

2010281734

ISBN 978-0-06-076539-2 (pbk.)

11 12 13 14 15 ID/RRD 10 9 8 7 6 5 4 3 2 1

If heaven too had passions
even heaven would grow old.

—LI HO,
"A BRONZE IMMORTAL TAKES LEAVE OF HAN"

Contents

III

The Good Life

PART I

THE PHOENIX

I have
Immortal longings in me.
—WILLIAM SHAKESPEARE,
ANTONY AND CLEOPATRA

Chapter 1

IMMORTAL LONGINGS

Late August, late afternoon, cloudy-bright.

We'd taken a corner table at the Eagle, just inside the red door on Benet Street. From there, the tavern's windows looked across to the tower of St. Benet's Parish Church, the oldest tower in the town of Cambridge and the county of Cambridgeshire. The church's foundation stones were laid almost a thousand years ago, when England was ruled by King Canute, son of the Viking King Sweyn Forkbeard, distant descendant of Gorm the Old.

A tavern stood across from that church tower in the year 1353, with beer for three gallons a penny—with shops and markets up and down the street, then as now, and around the corner the spires of the University of Cambridge, pointing at the same cloudy English sky. During the reign of Queen Elizabeth the First, the tavern was called the Eagle and Child. Elizabethan scholars would have stared up at its gently swaying signboard and (gently swaying themselves) remembered the myth of Zeus, who swooped from the clouds in the shape of an eagle, caught a child named Ganymede, and flew him off to Mount Olympus to serve as the gods' cupbearer, one of the immortals.

We'd been talking for an hour or two. The Eagle had been almost empty when we sat down. Now from the courtyard and the barrooms beyond we could hear more and more voices rising, glasses clinking. In the year 1940, in one of those barrooms, young pilots of the Royal Air Force who could not be sure they would come back placed chairs on tables, stood on the chairs, raised their cigarette lighters, and wrote their names on the ceiling with the soot of the flames. In another barroom, in the year 1953, two young biologists at the university used to meet over ale when they finished work at the Cavendish Laboratories, a few minutes' stroll down the lane past the church. James Watson and Francis Crick were trying to solve the structure of DNA, and hoping (they were not yet quite sure) that they'd figured it out. "So," Watson confesses in his memoir *The Double Helix*, "I felt a bit queasy when Francis went winging into the Eagle to shout that we had found the secret of life."

The Eagle remembers the pilots, and Churchill's praise: "Never in the field of human conflict was so much owed by so many to so few." And in the DNA barroom, the present management has engraved a line from Watson's memoir on the panes of the glass door: "I enjoyed Francis Crick's words, even though they lacked the casual sense of understatement known to be the correct way to behave in Cambridge."

Before the year 1500, when the College of Corpus Christi and the Blessed Virgin Mary, which is the nearest college of the university, built a chapel of its own, many of the school's dons and scholars would have begun their days in the parish church and ended their days in the tavern.

In the church, the prayers of the ages: *For this corruptible must put on incorruption, and this mortal must put on immortality.*

In the tavern, the toasts of the ages: *May you enter heaven late! May you live a hundred years! May you always drink from a full glass!*

They prayed for long life in the pews and they proposed long life in the pub, being the same mortals from morning to night.

"When you start talkin' about five-hundred-year humans"—said Aubrey David Nicholas Jasper de Grey—"five-hundred-year humans, or one-*thousan'*-year humans, most members of the general public get a li'l bit *nervous.*"

Aubrey was enjoying his fourth pint of ale at the Eagle, with dinnertime still some distance away.

This was our farewell drink. I'd spent most of the summer in London, and quite a few hours in Cambridge, listening to Aubrey over pints of ale. I'd heard him predict five hundred years for us, I'd heard him give us a thousand years, he'd hinted about a million years. He'd foreseen the coming of this new age of man in fifty years, or even as swiftly as fifteen. Now, because this was goodbye, Aubrey was trying to summarize his views, and to convert me once and for all, and I couldn't turn the pages of my notebook fast enough to keep up with him. I kept raising my hand to stop him while I scribbled, and while I scribbled he drank.

Tap, tap, tap went Aubrey's glass against the table, according to the sober testimony of my voice recorder. I'd placed it on his side of the table, by the cascades of his enormously long brown beard. From there it picked up every word, slurred or not—along with each groan and mortal screech of chairlegs and barstools against floorboards, and the frequent moments when Aubrey refreshed his voice and set down his pint.

"I mean, you have to appreciate the scale of this," Aubrey said. "It never leaves my mind." Think of it, he said: one hundred thousand human beings die of the infirmities of old age every single day. "One hundred thousand lives! I'm at the spearhead of the most important endeavor humanity is engaged in. Not easy to do, even though I don't often show it," he said, looking off. His face was struck by the late cloudy light from the windows of the Eagle, like a gibbous moon, three parts bright and one part in shadow.

Tap.

At a table near ours, a few people from the university explained to a guest, "*Cheers!* It means, *Here's to your health.*" Their guest returned with his own toast in a language that sounded Central European. It meant, *To life!* In 1940, the airmen of the RAF defended London and bombed Berlin. Now a sign on the wall warned "No Smoking" in English, French, Spanish, Japanese, and German.

"I should probably expand on that. You know, cuz people occasionally ask me about it, you know, how I cope with the—the *responsibility*, if you like," Aubrey said, with a small, apologetic chuckle. "Basically I just feel that I've got to put things out of my mind and get on with it. I just don't think about it. This is my fourth beer, you may have noticed."

Pause. *Tap.*

"And that helps, quite honestly. I do not like to think about it."

Tap.

Gravely he looked off again into the distance, toward the windows on Benet Street, stroking his beard. I had the feeling of watching a stage performance that I'd seen before. Aubrey had to be forgiven if he lost track of the speeches that he'd already made to me. He was talking with so many people around the world that he could hardly be expected to keep track of which speech he made when. But I felt sure he'd made this particular speech to me in another tavern with exactly the same lonely haunted stare into the distance. Was it here at the Eagle, over Abbot Ale? At the Tabard Inn, in Washington, over Foggy Bottom Ale? The Live & Let Live, in Cambridge, over Nethergate Umbel? My memory was getting a bit hazy. Somewhere before, he'd shown me this same look of agony, his secret anguish presented for my private viewing, with just the same half-turned head, looking aside and one-quarter down, the same phase of the moon. And watching him stare out the window, I felt sure that he had made the same speech in the same way with that same tilt of the head to many others by now. I had a sense of the crowd gathering around him.

Aubrey had stepped into the role that seems to open up again and again, the role of the prophet or sage who declares that we do not have to die, that we can be among the saved if we will follow him to safety. The same character in every age—an immortal character who is reborn endlessly, who has probably appeared more than once right there in that very tavern, given its own longevity, and the power of our longings.

Friends of mine, distinguished biologists, were a bit shocked to hear that I was talking with Aubrey de Grey. One of them warned me that if I listened to Aubrey I would be making "a martyr out of a molehill." But I didn't see Aubrey as either of those things; and I didn't think he was mad, either. Of course, he did drink. He admitted that himself. He had a long beard—but if you were charitable, you could say he wore that as a badge of office, the way an old-fashioned doctor would wear a white coat and a stethoscope. He really was highly intelligent, and he knew his field. He published papers with good people. He organized conferences, and respectable biologists came, and afterward some of them sat with him in the Eagle, too, listening and arguing. All in all, Aubrey was a remarkable phenomenon, a complicated mix of old and new, preposterous and plausible, practical and paradoxical, neither fish nor fowl. You could dismiss him with a laugh, but you would be wrong. In all these ways he was not unlike the field itself.

"This is lives we're talking about! It's people's lives," cried Aubrey now. "We're talking about one hundred thousand people a day. I'm driven by everybody. I used to be driven by myself. Now I don't think about myself, except that I'm making so much difference that it's important I don't get assassinated or fall under a truck."

Tap.

"Ultimately the sheer numbers are what drive me now. I just have so much disgust for any excuses. The idea one could postulate utterly vapid sociological concerns as genuine challenges to saving thirty World Trade Centers a day—I just don't have any words to describe . . ."

I held up my hand and scribbled.

"I don't do this anymore to extend *my* life span," Aubrey said again. "Small sliver of my motivation. My motivation is: it's going to be sooner, based on what I'm doing now. And I don't give a damn whose lives. I don't give a flying fuck whose. One way or another, someone's going to benefit."

Through the windows of the Eagle, I watched the clouds part again above the tower of St. Benet's. The sun flared against the pub's windowpanes with the tawny light of late August. Caught in those light shafts, Aubrey's pale face and his long brown beard were lit from one side once more, now one half bright, the other half in shadow, the moon rushing through its phases.

He said, "I mean, I think it's inconceivable that people born even ten years ago will die of old age, in spite of our pitiful reluctance to hurry—because serendipity will get us there in the end. It's just a matter of what we can do to accelerate things."

Our dates are brief, and therefore we admire what is old, as Shakespeare observes in one of his death-defying sonnets. We are brief, and therefore we admire a stone tower, a storied tavern, a Greek myth, an antique rippled windowpane, almost anything that seems to have more time than we do.

There was a time, not so long ago, when what we wanted to deal with our brevity was grace: grace to accept what we could not avoid, old age and death; courage to accept or to defy in the spirit what we could not change in the flesh. That was our condition from time immemorial. In every generation we worked toward grace. On every island and continent, we hoped for the best.

Now we live in a new time, with a somewhat different sense of time. Our life expectancies are increasing by about two years per decade, or about five hours per day, according to the standard estimates of scientists who study human life spans. That is to say, for every day we live now, we are given the gift of another five hours to live later on. While time runs out today, time pours in tomorrow: It is almost, but not quite, like the gift of an afterlife.

We find it hard to appreciate the scale and the suddenness of our success. In the Stone Age, most human babies died before they had reached the age of one or two. Few lived long enough to grow a single gray hair. The average life expectancy of Stone Age babies was probably not much more than twenty, although the evidence is scarce and the estimates are controversial (most of the science of human life span is controversial). When the Roman Empire was at its height, in the first century of the first millennium (a time when legionnaires patrolled Castle Hill, above the River Cam), Roman life expectancy was only a few years better: about twenty-five years. During the Middle Ages, in the first century of the second millennium, the era of the founding of some of the world's first universities—Bologna, Oxford, and Cambridge—life expectancy was about thirty years. During the Renaissance, it was thirty-three years.

In our corner of the Eagle that afternoon, the prints that hung on the wall just above Aubrey's head showed two jolly drinkers with tankards raised. Those gentlemen's powdered wigs and red coats would place them in the time of King George the First, Second, or Third. Back then, the tavern on this spot was called not the Eagle but the Post House. Horse-drawn coaches came rumbling into the cobbled courtyard every day to deliver the mail. By the Eagle's courtyard gate, you can still see the markers that guided in the coachmen—the old stone posts. Life expectancy in Georgian days rose toward forty years in England, less in its thirteen colonies. "When we see men grow old and die at a certain time one after another, from century to century, we laugh at the elixir that promises to prolong life to a thousand years," said Samuel Johnson. In reality, men and women were growing older and dying just a bit later each century; but by so little that Johnson was right to laugh.

By 1900, in the most developed countries of the world, including England and the United States, life expectancy had crept up to forty-seven years. That's all that a baby born in 1900 could expect. But if those babies survived and thrived and became parents themselves, their own children could expect to live longer; and *their* children could expect to live longer yet. By the end of the twentieth century, babies could expect about seventy-six years. Throughout the twentieth century, life expectancy changed so fast that for the first time in history, people became aware of it as a phenomenon that was extending their life spans during their own lifetimes. During the twentieth century we gained almost thirty years, or about as much time as our species had gained before in the whole struggle of existence.

In other words, this is a good time to be a mortal. Life expectancy today is roughly eighty years for anyone in the world's developed countries. And life expectancy is still improving, which is why each day we live now we are given the gift of more time down the road. It's as if we're all driving on a highway that is still being built, and the roadbuilders are adding to it at a good rate. Our bodies haven't changed. We haven't evolved. A few generations is too brief a time for our life spans to have gained thirty years through evolution. It's only that our circumstances have gotten more comfortable. A field mouse in the wild lives about one year. The same mouse in the safety of a cage lives about three years. With our farms and supermarkets and reservoirs and thermostats, we have done for ourselves what we have done for a pet mouse. We have tripled the life expectancy that our ancestors enjoyed or suffered in the wild.

To be clear: Life expectancy is the average age that babies born in any given generation or any particular year can expect to reach. Maximum life span is the longest that any member of a species is known to have attained. It is by the measure of life expectancy that our success has been most spectacular, so far, because we have done so well at helping babies and little children survive the dangers of their first years. But we are also doing better at helping people in their later years. Certainly, there have been fortunate people throughout history—those who were protected by great genes, wealth, power, luck—who have lived to a ripe old age. The ancients also had ancients. Among the pharaohs, Ramses the Second is believed by Egyptologists to have lived beyond the age of ninety, possibly to one hundred. Among the ancient Hebrews, when King David composed his psalms in Jerusalem, about three thousand years ago, our maximum life was thought to be about eighty years. David wrote, "The days of our years are threescore years and ten, and if by reason of strength they be fourscore years, yet is their strength labor and sorrow; for it is soon cut off, and we fly away."

Those lines from the Psalms were translated for King James the First by a committee of scholars in Cambridge in the first years of the 1600s. Some of the king's translators probably enjoyed a pint at the Eagle and Child. The longest-lived among them was a mild, cheerful, good-natured man named Laurence Chaderton. Chaderton could still read without spectacles when he was very old—one hundred years old, assuming his own count of the days of his years is reliable. He died on November 13, 1640, at the age of 103.

So there were happy specimens of old age in ages past. But now that our lives are so comfortable and secure that most of us reach eighty, more and more outliers have the chance to live well beyond eighty, and beyond the ages of Ramses and Chaderton. The world's record holder to date, Jeanne Calment, of Arles, France, lived to the age of 122 years and four months. The length of her days was 44,724. That is about the age that God promised Adam and Eve after evicting them from the Garden: "My spirit will not contend in man forever, for he is mortal; his days will be a hundred and twenty years."

The study of longevity is now in an almost feverish state. Twenty years ago, not many biologists worked on the problem. The field was small. It seemed old. You might say the science of eternal youth was looking and feeling its age. Efforts to extend the human life span in any serious, deliberate way had gotten nowhere since the studies of the ancient Greeks and Babylonians; since the tomb-builders and tomb-robbers of Egypt; since the glory days of the Taoist deep breathers, extreme dieters, and sexual athletes of China ("He who is able to have coitus several tens of times in a single day and night without allowing his essence to escape will be cured of all maladies . . ."). But today the science of longevity is growing fast. Once more it is turbulent, and painfully confused. It feels young again. The faces of the biologists who argue at international meetings about where we are, where we are going, and what we can or should do when we arrive, really are getting younger, because many new people are joining the field.

Specialists in this field call themselves gerontologists. The word comes from the Greek root *geron*, which means *old man*, but that suggests a focus that is misleadingly narrow. While it's true that the problems that limit our life span are normally most visible and cruel when we are old, gerontologists care about much more than the last years of life. They want to understand the whole span. Pediatricians treat the young. Geriatricians treat the old. Gerontologists try to understand why our bodies change from youth to age, why we age at all—why we are mortal. The problem of longevity is a deep problem because to understand it well enough to do anything fundamental about it, you first have to answer the questions: What makes us mortal? Why do we die? Why do we get frail year by year and ever more likely to die? When does the decline start—at forty? At thirty? When sperm meets egg? And where does it start—in the cells that compose the fabric of our tissues? In the way the organs talk, or fail to talk, to each other? What is aging? This is one of the hardest problems in biology. It is even harder than explaining consciousness. No one has managed to explain consciousness yet, either, but for some time we've had the source narrowed to a zone above the neck.

As gerontologists do begin to locate and explore the sources of mortality, many of them feel an incredible excitement. It's true, of course, that every mortal reaches the end of the road eventually—somewhere around the age of one hundred twenty, even supercentenarians seem to come up against a wall, and most gerontologists accept that wall as our limit. But they have hopes that they can help more of us reach it, and alleviate some of the suffering of old age along the way. As we approach some kind of limit now, it seems likely to most gerontologists that to go much further with either our average life expectancy or our maximum life span we would require a breakthrough in their science, in their understanding of the wellsprings of mortality. Only if they can figure out what aging is and what to do to change its rate will human life span take another big jump. Most gerontologists do not expect to see that breakthrough in their lifetimes. One group of conservative, well-respected gerontologists has proposed that our goal should be to add another seven good years to the human span. A few of the most enthusiastic people in the field have begun to argue for much more. If they are right, then our descendants in another few generations may expect to live as long as Moses, who is said to have lived 120 years; Noah, who lived 950 years; or Methuselah, the oldest man in the Bible: "And all the days of Methuselah were nine hundred sixty and nine years: and he died."

Aubrey de Grey thinks there is no limit. He is convinced that we can double or triple our life span again and again, and so onward and upward. We can engineer as long a life span as we like, "even life for evermore" (Psalm 133). That's hardly the majority view in gerontology. On the other hand, the field is so splintered and spiky right now that it's hard to find a majority view. Gerontologists can't agree on a way to measure aging, or what they mean by aging. Because so much of the action takes place in the United Kingdom and the United States, they can't even agree on how to spell the problem under discussion: aging or ageing. They fight over definitions of longevity, health, life expectancy, life span, maximum life span. But even in this overheated moment, Aubrey is the most fervent of them all.

Aubrey David Nicholas Jasper de Grey was born in London. His mother was a bohemian artist in Chelsea. She gave him his extraordinary name and some of his extraordinarily great expectations. (He never met his father.) He attended the University of Cambridge at the college of Trinity Hall, where he learned to drink beer, write computer code, and punt on the Cam, which is one of the favorite sports of students in Cambridge. He stayed in town after graduation, writing code.

Aubrey is six feet tall and medievally thin and pale, in spite of all the ale. When he stands up, his beard reaches a surprising distance toward his waist. When he sits down, it pools in his lap. "I find it useful to look unusual," he told me once. He looks like Methuselah before the Flood. Father Time before his hair turned gray. Timothy Leary Unbound. The beard never changes in length because Aubrey is always worrying it away at the edges, twining strands of it around his long pale fingers, even twisting the whole thing into a rope and pressing it to his shoulder when he sups his soup or blows his nose. He is a compulsive debater for his cause, and the beard is one of his weapons. "When I am stroking it like this you know you are all right," he says, "but when I begin to twist it like *this* you know I am about to pounce." He's made it his personal mission to demonize the bad old days when the science of gerontology was forlorn and we were all trapped and confined in a mortal existence; and to herald the days soon to come, when we will live a thousand years or more.

In a student town like Cambridge, with his beard, jeans, and T-shirt, whizzing around on his old bicycle, or striding through the campus with his faintly belligerent lope, or punting on the Cam, it would be hard to guess his age just by looking at him. In fact, he was born in 1963. That makes him one of the last babies of the great baby boom, or one of the first babies of the next.

In 1990 he met an older woman, an American geneticist named Adelaide Carpenter. She was born in 1944, in the dark of the war years. They met at a wild party that he threw in Cambridge. At the time he was a young man who liked to throw wild parties; she was an established biologist who'd made her reputation early and had lost her way in her career. She joined Aubrey in Cambridge. They married, and soon afterward, Aubrey became fascinated by biology and began his quest for immortality.

Aubrey thinks of aging as a medical problem. Since we all have this problem and it is invariably fatal, he believes we should hit it as hard as we can. He's convinced that every one of us will join the quest as soon as we realize that there are no technical obstacles to the cure for aging that can't be overcome, at least in principle. Our bodies are molecular machines. As they run they make mistakes, or give off toxic wastes they can't quite manage to get rid of. The mistakes are tiny. The wastes are submicroscopic. If we are lucky and enterprising we may find that the conquest of aging may require nothing more than a series of cleanup projects. Our bodies are like houses and cars. What we have to do (Aubrey puts this more positively: *all* we have to do) is keep up with the cleaning and repairs.

If we looked after our bodies properly we would stay healthy year after year after year, until we finally misjudged our step off a curb and ran into that truck. We would no longer die of our years. That is, we'd be no more likely to die at the age of ninety or 290 than we had been at the age of twenty. We would achieve a kind of practical immortality. Aubrey prefers the term "the engineering of negligible senescence," the creation of human bodies that hardly age at all.

It's a very British approach, in a way, consonant with a certain brisk stiff-upper-lip approach to immortality. In matters of the heart and mind and spirit, avoid muddle. In matters of the body, avoid rubbish. In some ways, you might even say, what Aubrey is proposing to do for the body is what civilization has accomplished for public health at large. Life expectancy stayed so low for most of human history because so many babies died at birth, along with their mothers. Improvements in housing, heating, farming, public health, the construction of sewage systems, the washing of hands in hospitals, and, in the twentieth century, the discovery of antibiotics—all these things together transformed our life expectancy. Public hygiene in Cambridge was horrible back in 1353. Aubrey proposes we clean up our bodies the way we have learned to clean up our cities and towns.

From time to time that summer I'd reminded Aubrey that I was listening as a reporter, not a disciple, that I was talking to many other gerontologists, trying to get the whole picture. Aubrey said brilliant and incisive things about what he called his Strategies for the Engineering of Negligible Senescense, or SENS. He'd published his manifesto, "Time to Talk SENS," in the *Annals of the New York Academy of Sciences*, in 2002, with half a dozen coauthors, including some highly respected scientists. And Aubrey had published many papers since—he was incredibly prolific, he seemed to write as fast as he talked—and often those papers were coauthored by specialists at the top of their fields. It seemed clear to me that Aubrey was a gifted amateur and provocateur. He'd pulled together a great store of arguments that the conquest of aging is at least a good goal, more than half a century after Watson and Crick, and that as a goal it makes sense. But he was also riding out into combat against almost everyone in gerontology. And in fact soon after that summer almost everyone in gerontology really did wheel around on Aubrey in one of the most spectacular, almost theological controversies in science in recent memory. Twenty-eight of the field's leaders signed a broadside in which they tried, in effect, to excommunicate Aubrey de Grey. "Ageing research is a discipline that is only just emerging from a reputation for charlatanry," they wrote. What a shame to see journals and scientific meetings give space "to empty fantasies of immortality." The goal of a few more good years or even good decades of life might be reasonable, but Aubrey de Grey's scribblings about SENS, with his talk of five hundred years, a thousand years, a cure for aging, were like essays on Aladdin's lamp. "Only a few people didn't sign," says the gerontologist Jan Vijg, one of the abstainers. Another abstainer was Judith Campisi. Vijg and Campisi are both distinguished gerontologists with a special interest in cancer. They think the conquest of aging is as reasonable a goal as the conquest of cancer, diabetes, atherosclerosis, Alzheimer's, or any of the other killers that rise up to get us in old age. If we have a War on Cancer, why not a War on Aging?

Once, talking with Campisi about all the controversies in gerontology, I threw up my hands. Maybe I shouldn't write about gerontology at all, I said. It's too confusing. It's too soon! "Well," she said, "it's not solved. You're writing about a problem that is not solved. I mean, if you want to write about a problem that's solved then you can write about smallpox." However, she said, if you want to talk about how a field has been muddled by human longings and blunderings for thousands of years, and has matured, then this is the problem to look at, because this is arguably the oldest problem in science, and it has suddenly come of age. "And if it matures even to the point where the field of cancer is now," she said, "if it can get to the point where cancer is now, it has the potential to change the course of human history."

Although I often reminded Aubrey that I wasn't riding out to the jousts with him, he seemed to forget my warnings from one meeting to the next. On that last day at the Eagle, he talked as if we were both believers; and now that I was leaving he was talking extra-fast, trying to sum up the situation and the needs of the campaign. With more and more hubbub around us and more and more ale inside him, Aubrey really was getting hard to follow.

"A v'iety—va'iety—*variety* of opinions . . ."

Both his hands fiddled rapidly with his apocalyptic beard and mustachios, although I could tell that he was trying to speak slowly. When Aubrey was in his cups, I'd noticed, his words came out thick and bushy, as if his tongue were cramped by his mouth, or his lips were too big. His voice itself stayed clear and reedy as a clarinet, his arguments remained as clever as ever, but something seemed to happen to the words. Somewhere in the tangles of his beard, they got brushy and muffled and indefinably squiggly, like a glimpse of figures, a line of horsemen, advancing through the brambles and the trees of a forest.

He wanted me to understand the difficult political situation he faced. Not only did he make the public nervous. He terrified most of his senior colleagues. They thought he was a menace. They were afraid that he would turn politicians and taxpayers against them. "And the main reason tha' I'm fabulously *dangerous*," Aubrey said, "is that I talk about these long life spans. Which is going to scare peo'le off doing anything. They're goin' to say, 'Oh, no, no, no—let's not fund any gerontology at all!'"

Tap, tap, tap . . .

"I'm no' a diplomat, you know," Aubrey said, and paused for a swallow.

Tap!

"A political animal, but no diplomat."

Pause. *Tap!*

"I don' find it easy to compromise. I find it easier to find solutions—to fin' killer punches." Aubrey mimed a roundhouse right at the air—*ka-pow!*—and laughed a roguish laugh, grinning at me eye to eye, conspirator to conspirator, as if the two of us really were about to witness the defeat of old age and the conquest of death, the cosmic victories that the world has longed for ever since Adam and Eve lost Paradise.

Pause. *Clonk!*

"Fucking aitch!" Aubrey cried. He'd just drained his glass and glanced at his watch. "It's already quarter past five! I mean, it's fine, you know, it's fine—this is valuable time. I'm scheduled to be at home for dinner at half past six—so what I ought to do is try to delay that. Let me nip over to the bar—they have a phone at the bar—and see if I can—"

From a stool at the far end of the bar, an old codger kept staring over at our table. I thought that same man—or someone just like him—had stared that way before from that same bar stool, with just that same ruddy, wrecked, xenophobic amusement, that leer from the Mad Hatter's Tea Party. (Was it the Eagle? The Live & Let Live?)

Aubrey's talk was toxic and intoxicating. Here was the dream of the ages. And yet, in some ways, what an awful moment to be dreaming about it, with so many mortal humans alive already; with so much of the living world in ashes around us, or near the flames.

"Depend upon it, sir, when a man knows he is to be hanged in a fortnight, it concentrates his mind wonderfully," said Dr. Johnson. And when we are told that the sentence of death under which we all live may be lifted, it makes our minds expand wonderfully, as if we have lived all our lives in a state of compression, increasing concentration, like a bird that is being lifted slowly on a finger toward the roof of its cage, or like a human body that is compressing with age, drawn down by gravity. It is strange and novel even to consider for a moment the possibility of negligible senescence; to consider that aging really might have a cure—a cure that we would desire, that is, not the one cure that the world has known since the beginning of time, which is death.

Of course, some of the gerontologists were so excited by the possibilities that they were only partly sober. They went weaving around the hard consonants and the insoluble problems that loomed up in the middle of their sentences the way a drinker emerging from the Eagle will sometimes go dodging around the lampposts and the parking meters.

This is where the science of mortality can take you. You can sit in the House of Watson and Crick, more than half a century past the Secret of Life, and pop down the rabbit hole, where every twist and turn is Wonderland, where each view is curioser and curioser, until you wonder how in the world you will ever get out. You can cross over the river and descend into depths where mortals have wandered for a thousand thousand years, trying to solve the riddle, wanting to know for sure, longing to climb back up and see the stars.

Chapter 2

THE PROBLEM OF MORTALITY

Sitting in the Eagle that summer, watching Aubrey try to sell me the conquest of aging, I sometimes thought, *What a piece of work!* He seemed to have almost limitless energy. Even after a whole day of talk, he still acted out every phrase. "Hard to *know*," he would say, with fevered miming of deep thought, eyes darting hither and yon. Then he'd fix me with a piercing pointed stare, or bow forward steeply, bringing the upper portion of the beard alarmingly close to the open mouth of his pint. The lower portion of his beard was safely out of sight, and I could imagine it brushing the floor.

He was given to mind-dumps, as computer geeks call them, which means the tipping and dumping of his entire stock of ideas from his cranium directly into yours with the help of all those gestures—the monkey and the organ-grinder and the music all in one.

What a piece of work!

But then, who can be cool about the problem of mortality? There is a heat around this topic from which we can never insulate ourselves. We are all mortals. It's our own body heat we are feeling. We can never cast a cold eye on life and death no matter how we try.

I've followed this science off and on for a quarter of a century. I was still a young man the first time I went to the New York Public Library, on Fifth Avenue, to read a little about the quest and its long history. I sprinted up the stone stairs between the stone lions, Patience and Fortitude, and looked up *immortality* in the card catalog, and thumbed through an almost endless series of titles on time-yellowed cards. I can remember a mystical sensation, browsing the old titles in those marble halls, a feeling of joining mortal souls throughout the ages in the investigation of this possibility of possibilities—the cure of cures, the conquest of conquests, the outwitting of all the powers that be. It was in the vaulted Reading Room that I first became acquainted with the plans for immortality proposed by Francis Bacon four centuries before us, and by Roger Bacon almost four centuries before him, and by many others far, far back before that.

About 4,500 years ago, when the first pyramids were going up, an Egyptian physician composed the world's first known medical text, beginning with injuries of the head and working down. The original of that treatise is lost, but part of it survives because a second scribe began to copy it onto a papyrus scroll more than a thousand years later. This second scribe copied forty-eight of the case histories, reached the chest, and stopped there, with more than fifteen inches blank at the end of the scroll—leaving off "in the middle of a line, in the middle of a sentence, in the middle of a word," according to one frustrated Egyptologist.

On the back of the papyrus is a prescription (in sloppier hieroglyphics, an Egyptian scrawl) for antiwrinkle cream. The recipe calls for large quantities of some exotic fruit or nut that scholars can't identify. Its flesh should be "bruised and placed in the sun." Then, after a great deal of husking, winnowing, sifting, waiting, mixing, evaporating, drying, packing into jars, scooping out of jars, washing in the river, and drying in the sun, the stuff is to be ground with a mortar and pestle; boiled; jarred again; and transferred to a vase of costly stone. When smeared on the face, the cream will remove all signs of age. "Found effective myriads of times."

That is how it has always been with us. From the beginning, the hard work of the prolongation of life; and the dream that is always as close as the far side of the papyrus, the dream of eternal youth.

Across the Mediterranean Sea, at about the time the Egyptian doctor was writing the original of that medical treatise, there was a man called Gilgamesh, son of Lugalbanda, who ruled as the fifth king of Uruk, in Sumeria. Gilgamesh became the hero of what is now the world's oldest surviving epic. Again about a thousand years passed before a scribe made a copy of the story that has survived. In the epic, Gilgamesh loses a friend, his comrade in arms, Enkidu. He suffers nightmares in the desert, and he goes searching for the secret of triumph over death. An old man who has survived the Great Flood and been granted the secret of immortality offers to tell him the secret if Gilgamesh can keep his eyes open for one solid week. But Gilgamesh closes his eyes and dozes off. Then, out of pity, the old man tells Gilgamesh where to dive in the river for a weed that will make him young again. Gilgamesh dives and finds the weed, but a snake on the riverbank steals it from him when he gets to the shore. So twice Gilgamesh has the answer in his hands and each time he loses it. In the end, all he gets to keep is the tale.

The problem of mortality appears in every mythology as one of the prime facts, if not the prime fact, to be explained; and in the West we have often blamed ourselves. In Hebrew Scripture, almost the first act of our first ancestors is to overreach, to pluck the fruit of the Tree of Knowledge. This is our original sin, for which God threw Adam out of the garden, "lest he put forth his hand and take also of the tree of life, and eat, and live forever." As with the Epic of Gilgamesh, we know where this story is heading from the beginning. Adam's name in Hebrew means clay or earth; and as soon as he and Eve eat of the fruit, God condemns them: "You are dust and to dust you will return."

We had it in our hands! We had it in our mouths! We could taste it! And then we lost it forever.

The ancient Greeks told the story of Prometheus, the Titan whose name meant "Forethought." For Prometheus's disobedience in stealing fire for our ancestors, Zeus chained him to a cliff, and punished all of humanity with old age and death. The Greeks also told the cautionary story of Tithonus. According to some traditions, he was a brother of Ganymede, the child whom Zeus kidnapped and made cupbearer of the gods. Eos, the goddess of the dawn, fell in love with Tithonus and begged Zeus to make him immortal, too. But she forgot to specify that Zeus should keep Tithonus from declining as he lived forever. Her lover grew smaller and more shrunken in body and mind until he was nothing but a cricket or a grasshopper. In the end she put him in a little cage.

Immortals lived up on Mount Olympus, mortals lived down

here on the ground. A Titan like Prometheus might try to save us; a god might swoop down and kidnap one of us; but most of us were mortal and moribund and we had to accept our fate, as one great soldier reminds another in the *Iliad*:

> *Like the generations of leaves, the lives of mortal men*
> *Now the wind scatters the old leaves across the earth,*
> *Now the living timber bursts with new buds*
> *And spring comes round again. And so with men:*
> *As one generation comes to life, another dies away.*

A history of our ideas about the conquest of aging was published by Gerald Gruman in 1966, and it is still one of the standard works on the subject. Gruman divides thinkers on the problem of mortality into "prolongevists," who want to extend our life spans, and "apologists," who try to reconcile us to our fates. In the East, Gruman says, there were fewer apologists than in the West. About eighteen hundred years ago, the great Chinese alchemist Ko Hung wrote treatises on this subject. Ko Hung was known as Old Sober-Sides. He was a prominent politician who fell in love with the idea of science and the promise of immortality. Why *shouldn't* we try to conquer aging, he asked: "We may perhaps be unable to make up our minds to believe that our lives can be prolonged or that immortality can be obtained, but why are we reluctant to make a trial? If only a slight success should come out from this trial, gaining thereby only two or three centuries of life, would even this not be better than the early death of the masses?" Even a mere two or three extra centuries would be nice. Ko Hung seems to have been able to dismiss easily, breezily, the tradition that opposed the idea. "As to Wen-tzu, Chang-tzu and Kuan-ling Yin Hsi . . . the final word is not there at all. Sometimes they equate death and life, saying there is no difference. They consider life as hard labor and death as rest. . . . They are not worth bothering with."

But in the West the voices of the apologists have always had more power. In the West, to propose that we try to reason our way to eternal life is to break our most ancient taboos about the use and abuse of knowledge. Our prevailing view has been that the gods didn't mean us to have it, and we shouldn't want it. So immortalist after immortalist has had to struggle against the hostility and ridicule of his time. The immortalists were up against what Gruman calls "the rationales, and rationalizations, which tend to entangle the prolongevist in a net of fear, guilt and despair."

Western civilization's ambivalence was sealed by one of the founding documents of the Renaissance, written by Christopher Marlowe, who went to college around the corner from the Eagle and Child, at Corpus Christi. Marlowe got his B.A. in 1584 and his master's in 1587. Just out of school he wrote the play *Doctor Faustus*. The play was based on folktales and legends, stories that may have been inspired originally by a petty fraud who sold a chaplain in jail a facial-hair remover. The ointment worked but it removed the poor man's face with his hair. That con artist became a magnet for tall tales. The legends and their ambitions grew far beyond hair removal. They were immortalized by Marlowe late in 1588 or early in 1589 in *Doctor Faustus*, a passionately ambivalent portrait. On the one hand, it's a pride-goeth-before-a-fall story. On the other hand, it's a sympathetic portrait of the man and his dilemma, as if the new age, the modern age, were confessing and lamenting its own flaws.

When the play opens, Marlowe's Doctor Faustus is looking for something worthy of him. Act one, scene one, Faustus in his study. What should he do with his time—argue philosophy? he asks himself. No, not good enough. "A greater subject fitteth Faustus' wit."

Be a physician, Faustus, heap up gold,
And be eternized for some wondrous cure.

But no, that would not be enough, either. He could do so much in that line that whole cities would escape the plague, and a thousand desperate maladies would be eased, and he would be remembered forever; and still he would want more, still he would not have reached high enough. Even as a great doctor, the greatest in the world, Faustus would still be mortal: "Yet art thou still but Faustus and a man."

There was only one way to achieve immortality in medicine, and that way was impossible.

Couldst thou make men to live eternally,
Or, being dead, raise them to life again,
Then this profession were to be esteem'd.
Physic, farewell.

Only the conquest of death would be worth the time of a scholar as gifted as Faustus. And Faustus knows that this is impossible. So he sells his soul to the devil for knowledge and power. In the end, the devil comes to take him down to hell.

Marlowe defined the apologists' position for modern times; and a few years later another Cambridge graduate, a young lawyer and philosopher, Francis Bacon, defined the opposing position. The grand project of Bacon's life was the reformation of learning. He had a vision of what it might mean to find things out; to go beyond all the learning and wisdom of the ancients and extend the boundaries of knowledge as far as the great voyages of exploration in his day were expanding the map of the world. Bacon's books helped launch the project we call modern science; and in a manuscript he called the *Valerius Terminus*, which he never published, he laid out his project's ultimate goal. "To speak plainly and clearly," he wrote, "it is a discovery of all operations and possibilities of operations from immortality (if it were possible) to the meanest mechanical practice."

When Bacon wrote those lines, King James the First had just taken the throne and ordered a new translation of the Holy Bible. Bacon argued that scholars should begin to read the natural world around them with the same reverence and care as they read Scripture. Natural philosophers in great teams, companies, and committees should begin to read and translate the great text of the world. Scholars in the universities of Cambridge, Oxford, Paris, and Bologna should go beyond their books. They should interview sailors about the Indies and Cadiz. They should interview miners about their quarries and blacksmiths about furnaces. Doctors should stop quoting Hippocrates and start examining their patients—which Renaissance doctors rarely did. Bacon thought they should even dissect the dead bodies of some of the patients they had lost and learn to decipher "*the secrecies of the passages*" (the italics are Sir Francis Bacon's). Then they might be able to figure out what went wrong. The traces of disaster must be in there in the body somewhere and doctors should talk about them. "Whereas now, upon opening of bodies, they are passed over slightly and in silence."

And this project was not sacrilegious but sacred, Bacon wrote. The first man and woman knew it all; they had "pure and uncorrupted natural knowledge," before they tasted the forbidden fruit and were thrown out of Paradise. Adam and Eve enjoyed not only perfect knowledge but perfect power; and there was no death in the world. Once we recover what our First Parents knew, we will conquer death again.

Most philosophers and theologians read Genesis rather differently. They explained our sorrows and our mortality with the doctrine of original sin. We sinned and we fell. Death was handed us by God because we deserved it—we asked for it. But Francis Bacon preferred the arguments of Roger Bacon, a thirteenth-century English friar, scholar, alchemist, and immortalist. (They weren't related, although their ideas were close kin.) In Roger Bacon's optimistic reading of the Holy Scriptures, God had not intended us to suffer and die. After all, God had made Adam and Eve wise enough to name all of creation. It was only after the Fall that they had lost their wisdom. To figure out what they had forgotten when they fell would be to recover Eden, and immortality.

In this telling of the story, what mattered most was not original sin, but original wisdom. We did not have to accept death. If we were smart, and worked very hard to learn the ways of nature, we could fight our way back to the original bliss of Adam and Eve before the Fall.

Francis Bacon titled his first proposal to King James *The Advancement of Learning.* In a famous passage he writes that no man "can search too far, or be too well studied in the book of God's word, or in the book of God's works." Much later, toward the end of his life, he laid out the research program for the quest for longevity in a little book entitled *The History of Life and Death, or, The Prolongation of Life.*

"To the Present and Future Ages," Bacon begins. "Greetings." He knows how heavily the West's long history of moral judgment will weigh on his readers, present and future; in the preface of his *History of Life and Death*, he acknowledges that his proposal runs against his readers' feelings of guilt and sin. He concedes that they may fear "that Knowledge hath in it somewhat of the serpent." They may remember the snake that tempted Eve and Adam to taste of the forbidden fruit of the Tree of Knowledge of Good and Evil, for which the Lord threw them out of Paradise. But the sin of the first man and woman was not the pursuit of knowledge, Bacon says. Their sin was pride.

Why don't we have medicines for long life? Bacon raises the question as impatiently and incredulously as the alchemist Ko Hung had done a dozen centuries before him. All we have is medicine to treat and cure diseases, Bacon says. "As for those things which tend properly to long life, there is but slight mention, and by the way only." He lists the recipes for long life that he has collected: potions of henbane, mandrake, hemlock, tobacco, nightshade, and dragonwort. The bulk of *The History of Life and Death* is a farrago of these recipes, like a hundred other how-to-live-forever books of its time and ours. Bacon even gives advice on underwear, and what we would call red long johns: "Some report that they have found great benefit in the conservation of their health, by wearing scarlet waistcoats next to their skin, and under their shirts, as well down to the nether parts as on the upper." The book includes potions for morning, noon, and night. Wine in which gold has been quenched. (That idea goes back to the alchemists—drink gold and become immortal as gold.) Pearls or emeralds, powdered extremely fine, and well stirred in the juice of four fresh lemons. Ivory, ambergris, horn of unicorn, the bone of the stag's heart, powdered.

Bacon offers all these potions only as suggestions, he says, although he has some faith in the "impregnation of the blood" with pearls and sandalwood and pulverized gold leaf. He also approves of "all those things which yield an odor somewhat earthy, like the smell of earth, pure and good, newly digged or turned up," including strawberry leaves, and strawberries, raw cucumbers, "vine leaves, and buds, also violets." He feels that "the smell of new and pure earth, taken either by following the plough, or by digging, or by weeding, excellently refresheth the spirits." (The English garden was already blooming.) "Nay, and we know a certain great lord who lived long, that had every morning, immediately after sleep, a clod of fresh earth laid in a fair napkin under his nose, that he might take the smell thereof."

Hope itself is good for the prolongation of life. Even a small amount, and we have a whole new view from a winter window, "so that hope appears to be a kind of leaf-joy, which may be spread out over a vast surface like gold."

What is radical and original about Bacon's book is not the prescriptions but the premise: that we should mount a vast testing program to search for the secrets of the prolongation of life; and that this program would become the triumph and centerpiece of the advancement of learning. The conquest of aging might not be possible in his own time, but it was not impossible, in his view; "those things are to be held possible," he wrote, "which may be done in the succession of ages, though not within the hourglass of one man's life."

After his death, Bacon became a hero of the scientific revolution. Generations of Bacon's followers read and reread the story of his last days, as told in John Aubrey's *Brief Lives*, which he wrote in the late 1600s: "As he was taking the air, in a coach with Dr. Witherborne (A Skotchman, physician to the King) towards Highgate, snow lay on the ground, and it came into my Lord's thoughts, why flesh might not be preserved in snow, as in salt. They were resolved they would try the experiment presently. They alighted out of the coach and went into a poor woman's house at the bottom of Highgate Hill, and bought a hen, and made the woman exenterate it, and then stuffed the body with snow, and my Lord did help to do it himself."

Assuming the story is true, that was one of the first experiments that Bacon ever tried with his own hands. It was also his last experiment. The snow so chilled him that he fell sick and died within three days. But the hen may have been, at least for a little while, preserved.

What we call modern science was born with these immortal longings. To many philosophers of the Enlightenment, the idea that we have to submit to aging seemed ludicrous. In England, Thomas

Hobbes implies the importance of life extension in his single most famous line. Hobbes had been one of Bacon's favorite secretaries; it was Hobbes who told Bacon's biographer the story of the fatal chicken in the snow. Death terrified Hobbes. He once wrote that his mother had given birth to twins: "to myself and to fear." Hobbes enshrines the Enlightenment's greatest hope for civilization by its negative in his description of life before civilization. He says the life of man in a state of nature was "solitary, poor, nasty, brutish, and short." The power of the line is in its emphasis on the last dreadful word.

The greatest French philosopher, René Descartes, was just as convinced as Bacon that mortals can solve the problem of mortality. Descartes turned from more abstract philosophy to the search for eternal youth when his hair turned gray at the age of forty-one. He wrote, "We could be free of an infinitude of maladies both of body and mind, and even possibly of the infirmities of age, if we had sufficient knowledge of their causes." He thought he had found the path himself, but he went down as ignominiously as Bacon. Descartes caught a cold in midwinter on a visit to Sweden and died at fifty-four. (Those immortalists didn't button up their coats.) One of his friends and admirers claimed that if only Descartes had not caught that cold, he might have lived five hundred years.

Many of the founders of modern science continued to hope that they or their followers would solve the problem of mortality. Who knew what wonders could be accomplished within a few centuries by the new natural philosophy? In 1780, Benjamin Franklin wrote in a letter to Joseph Priestley, the discoverer of oxygen, "We may learn to deprive large masses of their gravity, and give them absolute levity, for the sake of easy transport. Agriculture may diminish its labor and double its produce; all diseases may by sure means be prevented or cured, not excepting even that of old age, and our lives lengthened at pleasure even beyond the antediluvian standard. . . ." Beyond the standard of Moses, Noah, and Methuselah; or Seth, who lived to the age of 912; Enos, 905; Mahalaled, 895; and Jared, 962. "People that will live a long life and drink to the Bottom of the Cup expect to meet with some of the Dregs," Franklin wrote in a letter eight years later, trying to be philosophical about what he called his three incurable maladies, "the Gout, the Stone, and Old Age." But Franklin did not expect that people would have to meet and drink those dregs forever.

In France, another friend of Franklin's, the Marquis de Condorcet, one of the greatest minds of the Enlightenment, and a devout admirer of Francis Bacon, predicted that "a period must one day arrive when death will be nothing more than the effect of extraordinary accidents," when "the duration between the birth of man and his decay will have no assignable limit." Condorcet saw immortality as the climax of the advancement of learning. He wrote this passage of prophecy in a *Sketch for a Historical Picture of the Progress of the Human Mind* when he was in hiding from the Terror during the French Revolution. Soon after, Condorcet was captured, and died in one of the prisons of the Revolution at the age of fifty.

The founders of modernity felt so purely sunny and wholehearted about its future in its first days, when they could wave the dreams of immortality and patriotism together like a pair of flags. Thomas Jefferson commissioned a portrait of Bacon to hang in Monticello. Enlightenment optimism appealed enormously to the American founding fathers. So much of the local culture of self-improvement uncurled from that first seed. "Knowledge is power," said Bacon; and the chief value of power would be to buy us time. "Time is money," said Franklin. He knew that we would always want time as much as money, love, fame, or any other prize we hope to win in this life, if time allows. We would use our science to buy time. Once when Franklin was visiting England, he had a cask of Madeira wine shipped to him from Virginia. He found three flies floating in the cask. They made him think of death, and the future of the great experiments he had helped to launch. Franklin wrote to a friend in Paris:

> *I wish it were possible . . . to invent a method of embalming drowned persons, in such a manner that they may be recalled to life at any period, however distant; for having a very ardent desire to see and observe the state of America a hundred years hence, I should prefer to any ordinary death, the being immersed in a cask of Madeira wine, with a few friends, till that time, to be then recalled to life by the solar warmth of my dear country!*

Mixing, in one barrel, his hopes for science, immortality, and the United States of America.

Mortality seems to be a problem for which every people on Earth began reporting solutions the moment they invented writing. And yet each generation is ready to believe the problem is solved or about to be solved at last. These overenthusiastic reports right from the beginning are like telltales, the ribbons that sailors tie to the tops of their masts to show which way the wind is blowing. Our perpetual readiness to believe we have the answer is a measure of the force of our private hopes and our ambitions as a civilization.

The great Russian biologist Elie Metchnikoff took up the problem of mortality in 1914 with the complaint that "science knows very little about old age and death." He developed a theory that we are slowly poisoned to death by the bacteria in our bowels. As a remedy, he drank sour milk every day. He explained his theory in a popular book, *The Prolongation of Life: Optimistic Studies.* Not long after Metchnikoff's death, two biochemists at the Rockefeller Institute for Medical Research, in New York, showed that fruit flies bred without bacteria in their guts live a shorter time than fruit flies with bacteria in their guts. Public fascination with Metchnikoff evaporated, but yogurt never went away.

At about the same time there was a craze for grafting monkey testicles onto old men. It built on the work of a French neurologist and physiologist in the late nineteenth century, a man with the euphonious name of Charles-Édouard Brown-Séquard, who was Harvard's first professor of the pathology of the nervous system, and who coined the word "rejuvenation." Brown-Séquard gave himself shots of fluid that he'd extracted from the testicles of young dogs and guinea pigs. He was seventy-two at the time, but he looked at least twenty years younger, and he claimed that the injections restored some of the sexual potency of his youth. He announced his experiment in a famous lecture in which he told his audience that he had only just that morning "paid a visit" to his young wife Madame Brown-Séquard. With that "paid a visit," which has a double meaning in French, Brown-Séquard created an international sensation. He inspired a series of doctors to take up rejuvenation, including Eugen Steinach in Austria and Serge Voronoff in Russia, who became celebrities themselves. It was Voronoff who specialized in transplants of ape testicles, beginning with his first operation in 1920. They were expensive procedures but three hundred men are said to have undergone them in the next five years. Steinach was nominated half a dozen times for a Nobel Prize for his rejuvenation operations (although he never won). He didn't do transplants; he did vasectomies. The hope there, as a historian of medicine, Diana Wyndham, explains, "was that, instead of giving life to children, aging men would give life to themselves." Steinach was even more celebrated in his day than Brown-Séquard and Voronoff, thanks in part to his book *Rejuvenation through the Experimental Revitalization of the Aging Puberty Gland*, which he published in 1920, the year that he began offering the operation. Men who got the vasectomy were said to have been "Steinached." An article in *Scientific American Monthly* reported that year, "It seems that the magic hand of science has found that Elixir of Life . . . for which Faust bartered his soul." According to the article, old men who had been Steinached "not only looked fresher and young, but felt an increase in strength and vigor, while aged trembling hands grew steady, feeble tottering steps became firm and failing masculine instincts and impulses acquired new vitality." In the 1920s, more than one hundred Viennese university professors and teachers were said to have been Steinached, including Sigmund Freud. Freud may or may not have felt younger after being Steinached. He didn't like to talk about it.

Another famous man to be Steinached was William Butler Yeats, who read about the rejuvenation operation in a popular book called *The Conquest of Old Age*. Yeats underwent the operation in the spring of 1934, and his surgeon continued the experiment afterward by inviting Yeats and a beautiful young poet to dinner at his house. Yeats was sixty-nine. His wife was forty-two. The young poet, Ethel Mannin, was thirty-four. They may have had an affair. Yeats did start an affair with another young poet and actress, Margot Ruddock, six months later. Around Dublin people began calling him "the gland old man." Yeats believed the operation had made him a new man. Only a month before his death he wrote, "I am happy, and I think full of an energy, of an energy I had despaired of." The operation, or faith in the operation, seems to have helped give Yeats an outpouring of great poetry and wild romance in the last years of his life.

Eventually the world forgot Steinach, but kept the poetry.

Of course, it's easy to laugh at these early attempts at rejuvenation. Sympathetic historians of aging remind us that these were sophisticated and reasonable projects for their time. Brown-Séquard is now recognized as the founder of the science of endocrinology and the study of sex hormones. Metchnikoff won the Nobel Prize in 1908 for research in immunology. And it was Metchnikoff who coined the word "gerontology."

One of the most distinguished biologists who tried to solve the problem of mortality in the early twentieth century was Alexis Carrel, who won the Nobel Prize in 1912 for his pioneering work in vascular surgery. Carrel was a star researcher at the Rockefeller Institute. His labs and surgery rooms filled the whole top floor and the attic of Founder's Hall, which was the first building on the campus. The fifth floor was given over to his labs, and his operating rooms were in the attic. Carrel had them all painted black. The walls, the furniture, and every piece of equipment in his operating rooms had to be black, and everyone who worked in them had to wear black surgical masks and gowns. (Whatever else it did for the experiments, the black added drama.) Carrel had developed an early antiseptic, as a surgeon in the French army during World War I. Then he turned to, among other things, the project of breeding mice for longevity. He was a short, odd-looking man, one eye brown, one blue, and he spoke with a heavy French accent. He got extraordinary attention in the papers for his claims that he was keeping cells alive year after year in a petri dish. Carrel wrote in 1911 in the *Journal of the American Medical Association* that his results "demonstrate . . . that death is not a necessary, but merely a contingent, phenomenon." Carrel thought it might soon be possible to keep a human head alive as long as he wanted. Like Brown-Séquard, Voronoff, and Steinach, he made headlines: "Carrel's New Miracle Points Way to Avert Old Age," trumpeted the *New York Times*. "Flesh That Is Immortal," shouted the *World's Week*.

But Carrel's immortality project turned out to be premature, too. The cells in his petri dishes were not immortal after all; they were being replenished with young cells month after month by the young scientists and assistants who tended them in the black attic, wearing their black gowns. No one knows if the sorcerer's apprentices deceived the great man deliberately, or if it was an honest mistake. Carrel's claim was proved wrong only in 1961, when Leonard Hayflick, a cell biologist at the Wistar Institute in Philadelphia, demonstrated that normal human cells do not divide indefinitely in a petri dish. Hayflick was able to show that in the open air (which is 21 percent oxygen) our cells divide about fifty times. In air with the reduced level of oxygen that prevails inside our bodies (about 3 percent) human cells in a petri dish will divide about seventy times. Then the cells get old and tired, a state that is known to cell biologists as senescence. The cells in Carrel's dish went from symbols of immortality to symbols of mortality. Hayflick has spent the rest of his career (he is now eighty-one) arguing that it is impossible for gerontologists to extend the human life span.

Our long lives now are made up of all the extra seconds, minutes, hours, and years that the advancement of learning has given us. These gifts of time come down to us from even the least and meanest inventions, from the privy, the chamber pot, the mousetrap, the pitchfork and scythe, the broom and dustpan, from every little lifesaving or timesaving gadget ever invented, including the first nail and the first screw. Benjamin Franklin contributed more than a few of those inventions, including the lightning rod, the Franklin stove, and Franklin bifocals. With the bifocals, he was tickled that he had solved at least one of the problems of old age. He wrote to a friend that "if all the other Defects and Infirmities were as easily and cheaply remedied, it would be worth while for Friends to live a good deal longer." But they've all contributed to the same project, from the first fire, and the first chipped, serrated flints; from the first look our ancestors took at the horizon when they stood upright on hind legs; to those bifocals.

This is just what Bacon hoped for the advancement of learning. The goal and index of all our learning would be long life. And Bacon's prediction did come true. Each of us lives on the capstone of a pyramid that is the sum of the contributions of every invention and material improvement since the Stone Age. Our bodies survive for seventy or eighty years because of every advance in medical knowledge from the most basic discoveries in anatomy to the latest discoveries of Baconian "secrecies of the passages." The hipbone is connected to the thighbone. The sacrum is connected to the iliac. (And the joint that connects them is the sacroiliac.) Scientific revolution and technological transformation have changed human life for better and for worse in innumerable ways, and the greatest change, the sum of it all, the ultimate payoff, is indexed by our life spans. This is the biggest accomplishment of our species to date: it is the cumulative gift of fire, language, science, art, law, and medicine. This is the apex of the pyramid that we've been constructing from the beginning. The building of this Rome took the whole of human history, and all the roads led here.

Better food, better water. Better public sanitation and personal hygiene. Better education, also known as the advancement of learning. You can frame all of this as a quest to extend our lives, just as the founders of the Enlightenment hoped. The failures have been staggering. But the progresss of our civilization has made life, on the whole, less nasty, less brutish—and certainly longer. And even though the quest for immortality keeps falling short of its ultimate goal, generation after generation, it does succeed in the most practical, pragmatic, basic ways in the task of sustaining each generation for a little longer than the year before. That thought goes back way before Bacon, too. According to Sumerian legend, Gilgamesh was the first to dig wells. At least one Babylonian tablet seems to suggest that Sumerians invoked his name whenever they started to dig a well. They chanted, "Well of Gilgamesh!"

Chapter 3

LIFE AND DEATH OF A CELL

Ever since he was a boy, Aubrey de Grey has felt that he was cut out for an extraordinary destiny. He was always interested in making a difference in the world by doing things that were hard, things that required a certain virtuosity, things that the world thought well-nigh impossible. It started with playing piano. His mother wanted him to play, but he wasn't particularly apt. He came to the conclusion early on that it was a waste of his time. If he was out to contribute something unique to the world, to change the world, this was not his way: With lots of people good at it, much better than he would ever be, why did the world need another piano player? He still sees no point in doing things that other people are doing well already. "That's certainly one of the reasons why I don't have kids—one of the main reasons," he says. "Anyone can have kids. A lot of people are very good at it. I want to make a *difference*." Nor does he claim to understand people who work in highly crowded fields of science, fields in which, if they went up in a puff of smoke, the very same thing they'd done would be achieved by someone else five minutes later. "That's completely incomprehensible to me."

He has a sort of roguish strut. His arrogance is made just tolerable by his matey readiness to confess it, and his willingness to try and fail at great things. At Cambridge it is traditional to race around the Great Court of Trinity College. A student named David George Burghley once ran all the way around while the Trinity clock chimed one, two, three . . . to twenty-four, a sprint of 373 yards in 43 seconds. Burghley later became an Olympian and the hero of the movie *Chariots of Fire*. Once, after a party, when Aubrey was a student at Cambridge, he ran that race against the Trinity clock, trying in a way to go Burghley one better. Aubrey sprinted around the Great Court stark naked. He was happy to get more than two-thirds of the way around before the bells tolled twenty-four. But he slipped on some cobblestones along the way and fell on his face. The next day he hobbled around Cambridge with a magnificent pair of purple-and-black eggplant-colored circles around his eyes. That adventure earned him the nickname Aubrey Aubergine, after a sad-eyed character in a British series of children's books, part of a gang of unwanted fruits, vegetables, and nuts. "Aubergine" is not only the color of eggplants, it is also computer jargon: "A secret term used to refer to computers in the presence of computerphobic third parties," according to the Free On-line Dictionary of Computing.

Aubrey studied computer science at Trinity Hall, Cambridge. When he graduated, in June 1985, he was hired by Sinclair Research, a high-tech company in town. There he got involved almost immediately in a suitably difficult artificial-intelligence project. A computer program involves a long sequence of commands written in code. If even one line contains a glitch the whole program can crash. Aubrey and a friend at Sinclair, a software engineer named

Aaron Turner, began working on the design of a program that could inspect any other program on Earth and debug it automatically. The creation of such a Cure-All is one of the legendary problems of computer science. Its solution is believed by most programmers to be virtually impossible.

While Aubrey was working on the Cure-All, Adelaide Carpenter arrived in Cambridge on a sabbatical. She was a professor at the University of California, in San Diego, at the time. It was her second sabbatical and she was burned out and thinking of quitting science. One day she heard about a party from a graduate student in her laboratory, a young man from whom she bummed cigarettes. Every day she bummed a cigarette, and every week she bought him a pack.

It was a birthday party for someone young, and she did not expect to enjoy herself. As she tells the story, she was standing nervously in the front room when a handsome young bloke came up to her and said, "Justify your existence!" Almost immediately he was called away to deal with something electrical. Apparently it was the bloke's place. She watched him go. She thought he was charming and cocky. He was twenty years younger, the age of her students. He walked with the air of a man who owns every place he goes, a man who is always saying: Remember my name. She lost him for a time, but later on they bumped into each other again by the wine. Then they danced at each other for a while. There was a couch. Then his bedroom was free.

The morning presented an unusual situation for her. It was awkward. She realized that they did not know each other's names.

His name was Aubrey David Nicholas Jasper de Grey.

* * *

It was after they were married, sitting across the table from each other at breakfast and dinner, that Aubrey began to quiz Adelaide about the biology of longevity. He gathered that almost no biologist was working on it. Whenever he brought it up with Adelaide, she told him that aging was almost impossible for biologists to study and absolutely impossible for doctors to treat or to cure. "I know now that most biologists around that time did take that view," Aubrey told me. "They looked very much down on gerontologists." He assumed that the problem was too hard to work on. When scientists could work on it, they would.

Like anyone else, Aubrey says, he was interested in the problem of aging, and of course he loved a problem with the reputation for impossibility, but Adelaide did not seem to want to talk about it. She was happy to tell him about her experiments over the breakfast or dinner table, but she seemed sadly pessimistic about the subject of old age. She herself studied the brighter side of life. She worked on meiosis, which is part of a series of magic acts by which cells make sperm and eggs. All the action at the start of life, from meiosis to the meeting of those sperm and eggs to the growth and development of embryos, is magic on the finest scale. By the time Aubrey began learning from Adelaide, the science of development was one of the most successful fields of study on the planet. Using the tools that Watson and Crick had discovered in the double helix, more and more biologists had devoted themselves to the study of that spectacularly orderly sequence by which one cell becomes two, two become four, four eight, and thence to the embryo, and thence to the newborn baby. Adelaide had made a name in the field by discovering a tiny molecular mechanism she called a recombination nodule. It lies at the very beginning of all this spectacularly orderly growth. She studied the phenomenon in fruit flies, which are convenient animals to watch in embryo. They develop very quickly, not in nine months but in eleven days, they're cheap to raise, and their embryos have much of the same machinery as we do. It was a small but extraordinarily successful field, the study of the developing fruit-fly embryo; three of its pioneers, Ed Lewis, Christiane Nüsslein-Volhard, and Eric Wieschaus, later shared a Nobel Prize.

Aubrey couldn't see why we shouldn't understand the second half of life with the same success as the first half. But Adelaide had grown up and come of age in a time when the study of the first half of life was booming and the study of the second half was stagnant. And development is beautiful anywhere in life you look. The development of the roundworm *Caenorhabditis elegans*, for instance, is spelled out in such fantastically careful detail that when it crawls to maturity it will have precisely 959 cells in its body, not one cell more and not one less. That was marvelous—elegant—to a developmental biologist. The subject of aging was depressing, with so much living order in the beginning and so much fatal disorder at the end. It was easy to see that you could hope to find something like a recombination nodule. It was hard to see how you could hope to find a destruction nodule. If there were destruction nodules they worked very differently. They didn't shuffle the deck the way sperm and egg shuffle genes. They tossed the deck into the air, or threw the cards out the window one by one, or set the deck on fire, or lost it under the hedge in the rain. And where would you look for the secret? With development you could look at almost any embryo and find the same machinery at work. With aging you were talking about the whole life span, and the life cycles of living things are bizarrely distinct and differentiated one from the next. There is so much variation in aging that it is hard to know where to hunt for first principles and first causes. Where is the speck of dirt that will turn out to matter?

By and large, of course, in the big picture, the pattern on display is the same everywhere. But in the details, there is infinite variation. Back in the early 1600s, in his program proposal *The History of Life and Death*, Bacon had recommended that "the higher physicians" begin by collecting life histories more or less at random and then looking for patterns. Even in the late 1980s, biologists interested in aging were still doing that. They now had many theories, and they had many more volumes filled with observations, but most gerontologists were still lost in the forest—in what Bacon called the *Sylva Sylvarum*, the Forest of Forests. Where exactly should you look? Imagine grappling with the rock at the bottom of a cliff. You need a purchase on the cliff face. If the face is all smooth you can't start up. Or if the base is all clinking rubble it's hard to start up. The study of aging suffered from too many barriers, too many theories, too many observations.

Aging is not neat, and of course it is the neat patterns that are the simplest for science to solve.

It is amazing to watch the development of an embryo from the meeting of a sperm and an egg. First you have that one fertilized cell. From that one you get two, then four, then eight. By the arithmetical law of doubling you soon arrive at the astronomical number of cells in an early embryo. That much is astonishing but basically comprehensible because it is pure arithmetic. Then new kinds of order begin to emerge where there had been nothing but that little soccer ball. You get the head and the tail and the gut and so on. And of course the orderly progress of development doesn't stop there. If the sperm and egg came from frogs, you get a tadpole, which metamorphoses into another adult frog. If the cells came from a painted lady butterfly, you get a caterpillar, then a chrysalis, then a painted lady. All of that is regular, repeatable, and predictable, again and again. If it weren't, you wouldn't have frogs and butterflies. And likewise with people. Frogs, butterflies, and human beings all begin in very much the same way. Then come spectacular divergences, all of which are predictable, all of which are in some sense as scripted and lovely as Shakespeare's sonnets. The rise of each organism from that fertilized egg is one of the most beautiful things on Earth to watch. You can watch again and again and feel the same sense of wonder as the very first time you saw it. And part of the beauty is in the predictability: the sonnet unfolds the same way with each rereading.

But then of course comes the decline. If you look closely at aging organisms you see endless, desperately depressing, unpredictable variations on the theme of decline. It is nothing like the detailed harmonious unfolding of the beginning. It is more like the random crumpling of what had been neatly folded origami, or the erosion of stone. The withering of the roses in the bowl is as drunken and disorderly as their blossoming was regular and precise. In growth you see the genius of life, and in its slow destruction you see chaos.

Think of the creation of a work of art and its destruction. Leonardo made many notes in his secret notebooks about his search for durable pigments that would keep their colors. But of course they faded and there is no artistry visible, no genius apparently at work in the fading of the *Last Supper.* Michelangelo once made a snow sculpture in the garden of his patron Lorenzo de' Medici. He took that commission with reluctance, to indulge his patron, who was the most powerful man in Italy. What happened to that snow sculpture within a few days was nothing like the act of genius that went into the shaping of it. The making and the melting were two different processes entirely.

Of course, there are a few regularities in aging. Seen from a certain distance, in fact, our rise and fall are so predictable that we can usually guess a stranger's age within a few years. Shakespeare wrote the best description of the problem of aging in the famous speech in *As You Like It,* "The Seven Ages of Man." "All the world's a stage . . ." And each of us plays seven parts. First the infant, mewling and puking in his nurse's arms. Then the whining schoolboy who creeps unwillingly to school. Then the lover who sighs his way through a poem to his mistress's eyebrow. And so on.

> *. . . Last scene of all,*
> *That ends this strange eventful history,*
> *Is second childishness and mere oblivion;*
> *Sans teeth, sans eyes, sans taste, sans everything.*

That description of aging in *As You Like It* is one of the greatest passages in English. It is characteristic Shakespeare, intimately sympathetic and cosmically amused. Shakespeare seems to feel each age and stage as if he has already lived it himself; at the same time he views them all from outside, as if he is looking down at the whole theater of life from ten thousand feet up in the sky. And we can still recognize them now and they all move along pretty much on the same schedule. In that sense our decline does seem to be orderly. Even centuries later we all recognize Shakespeare's seven ages of man. So many of the little painful moments of the decline seem to progress on schedule. You complain to an older friend at dinner that you can't read the print on the menu anymore. How old are you, he asks, forty? Yep, right on schedule. A man still reaches the age of spectacles on nose; and then the age when his big manly voice begins to pipe and squeak once more, "turning again toward childish treble."

So aging is both predictable and unpredictable. It is both inevitable and erratic, even in families. One sister gets bad knees at thirty-five and can't jog anymore. Another sister gets a bad back at sixty and can hardly walk. The third sister gets such a lucky body that she is still running marathons at seventy-five. If you had that kind of wide variation in the womb, none of those girls would ever have been born. The very broad pattern of aging is the same for each of us, always the same slow subtraction of powers. But where and when each power gets subtracted—that seems almost unbearably random from one body to the next, or even from one part of the body to the next. The Koreans have a saying, "Each finger can suffer." Every people has a saying that translates, roughly: We all have to go sometime. But which part of you will go first, and which next, or how, when, and where you will suffer, no one can say.

This chaos makes it hard for biologists to figure out what is going on in aging bodies, or where to try to intervene. When you are looking at order, you can investigate its causes and hope to understand them. You can hope to find processes as neat and clear as the progression itself, as for instance that arithmetic code, two, four, eight, sixteen. . . . But there doesn't seem to be anything so arithmetically predictable about aging except that it happens again and again, happens every time. You watch, as Shakespeare watched,

And every fair from fair sometime declines
By chance, or nature's changing course untrimmed.

Every beautiful human body loses its beauty and then loses its life, by chance or by something built into the nature of the body itself. But no one can say what that something is or where it might be hiding. Where is it? What is it? For years the murkiness and muddiness of the biology of aging has scared away many of the best and the brightest.

Even if you step back and look at the problem of mortality in the tree of life as a whole, you see confusing variation. It's not enough to say that aging is the way of all flesh, because not all flesh does age. Life cycles are so diverse that you can find arguments for almost any theory of aging depending on which creature you study, which chapter and verse you choose to quote. Consider the hydra. Hydra is one animal that may be practically immortal. Biologists have argued for more than a hundred years: Does hydra age very, very slowly, or does it live forever?

Hydra is a sort of stick figure of life, a tiny tube with a head at the top and a foot at the bottom. Around its head and mouth, it has tentacles that it waves around like the arms of a squid or an octopus. Some species of hydra have just a few tentacles; other species have as many as a dozen, which can stretch four or five times the length of the body. From these tentacles they can throw a sort of harpoon on a thread, armed with neurotoxins, to paralyze their prey. The hydra is related to sea anemones, jellyfish, and the Portuguese man-of-war. They're found in most freshwater ponds and streams, and most of them are only a few millimeters long.

Our own bodies are vastly more complicated. Men have sperm cells, women have egg cells. We all have muscle cells, nerve cells, skin cells, liver cells—about two hundred different kinds of cells. But the stick figure of the hydra is made up of only about twenty kinds of cells, including cells of the outer layer of the skin, the inner layer, and nerve cells to control the waving and firing of the tentacles, and both sperm and eggs (each hydra being a hermaphrodite). Cells in the body column, the stick of the stick figure, are constantly making more cells—more copies of the hydra's twenty types. Some of these cells migrate up to the tips of the tentacles and then are sloughed off. Other cells migrate down to the tip of the foot and then are sloughed off. And some migrate into buds in the middle of the stick figure. There they break off and grow into new hydra.

Biologists who study the near-immortality of the hydra describe its body as a kind of fountain. The cells in the stalk are at the base, and the fountain sprays upward and outward in all directions, slowly.

A cell that forms in the center of the body, somewhere inside the long stick of the body column, takes about twenty days to reach the outer limits of the body, the head or the foot, and fall away.

Daniel Martínez, a biologist at Pomona College, in California, is studying a collection of hydras that was begun in the late twentieth century and so far show no signs of aging. The fountain of each hydra in his collection remains as vigorous as ever. Each hydra buds as much as ever. One individual hydra during its first four years of observation produced 448 buds that matured into full-grown hydras. At the same time, each hydra in that first four years of the study replaced its whole body sixty times over. Even in the lab, of course, an individual hydra does die now and then, if it is mishandled by a student or a technician. But otherwise these stick figures seem prepared to live forever, or at least a very very long time.

Some gerontologists wonder if the hydra really does come as close to immortality as Martínez claims. The case rests mostly on a single paper he published in 1998. Martínez hasn't published much on it since, and no one else has followed a population of hydras for nearly as long. Gerontology is such a young science that it is still full of blanks like that along the frontiers. Single travelers come back with reports that some choose to believe and others disbelieve. It is almost as bad as the days of the old writer Sir John Mandeville, in the fourteenth century, who describes the precise location of Paradise in his book of travels, and how hard it is to reach by rowing upriver because the currents are so strong.

Gerontologists do agree that the source of the fountain of the hydra is traceable to certain cells concealed in the central stick of the stick figure, the body column: stem cells. They are called stem cells because they are able to make all twenty varieties of cells of the simple body of the hydra; all of the diverse cells of the hydra's body can be said to stem from them. We, too, with our far more complicated body plans, have stem cells concealed in the interstices of our bodies. But our bodies do not replace themselves successfully and perpetually as the hydra seems able to do. So the question is: Why can't we do what the hydra does? Why can't we do what we ourselves seem to be able to do at the age of twelve, when we are still green and growing?

Of course hydras are simpler than humans. But simplicity alone can't explain the difference between the stick figure and the human figure, because there are creatures that look and act like hydras but are far, far simpler. In fact, they are almost as simple as life gets: they live their whole life cycle as single cells. Even so, like human beings, they do grow old and die.

I know this because it was one of the first things I wrote about as a science writer. I was just starting out in the early 1980s when I went to visit one of the then-rare researchers who was working on aging, an elderly biologist named Maria Rudzinska. She worked at Rockefeller, on York Avenue, in Manhattan, where Alexis Carrel had made all those black-draped, spooky efforts to cheat death with immortal cells at the turn of the twentieth century. I'd heard about Rudzinska's latest experiment and had wooed her for months with polite letters before she'd agreed to talk with me.

Rudzinska was late for our meeting, so I parked myself in a chair by the door to the Rockefeller cafeteria and opened a book. I'd been doing a lot of reading about the science of life, mortality, and longevity, and had discovered Bacon's *History of Life and Death*.

I was sitting there by the cafeteria door, scribbling notes in the back of my book, when an elderly voice called my name. I looked up and saw Maria Rudzinska. Her hair was gray, pulled back in a tight bun, her glasses thick and mended with tape. Behind the goggle lenses, her eyes looked huge and watery. She stooped. Her cardigan hung loosely from her shoulders, as if she had been wearing it ever since the age when it had fit. Around her neck she wore a medallion so big that I had to force myself not to stare at it, a big bronze sun. She was so stooped and the chain was so long that the sun hung down almost to her belt.

"Look, he is always writing!" she exclaimed, speaking not to me or to anyone standing nearby but to an invisible audience. I recognized that voice and that audience. They both belonged to the Old World, where people loved writers who were always writing, because they themselves were always reading.

Rudzinska led me into the cafeteria. Over lunch she told me her story. During the war, she said, she and her husband, Aleksander Witold Rudzinski, had fought in the Resistance in Warsaw. Aleksander had been wounded. At the same time, working under great difficulties, without supplies, sometimes without much food, she had managed to carry on her research. I've long since lost my notes, but as I remember the story now, she told me that she'd scraped gunk from the side of an aquarium tank in a half-abandoned laboratory and studied what she found through the microscope. It was unusual for a woman to become a scientist in those days; much less while half-starving in the war. But she'd been entranced by the lives of single-celled animals ever since her first scientific paper in Cracow in 1928: "The Influence of Alcohol on the Division Rate in *Paramecium caudatum.*"

Somewhere in the gunk on the wall of the tank she found a rare, single-celled pond creature called *Tokophrya*, and she fell in love with it. The adult *Tokophrya* looks like a miniature hydra. It's another stick figure of life. Its body is a stalk. At the base it has a sort of suction cup called a holdfast. At the top it has sixty or seventy tentacles. The tentacles stand out from it in long straight lines like rays around a child's drawing of the sun, waving in the water. If a paramecium swims too close, it gets stuck and impaled. Then *Tokophrya* sucks its victim's innards through the tentacles, as if it were drinking its prey (still alive and struggling) through a dozen straws.

In all that, it is like the hydra. But the way *Tokophrya* gives birth is more like us. A tiny bud, a baby, grows in the cell inside a miniature womb called a brood pouch. When the baby is ready to be born, it whirls and struggles in the pouch for ten or twenty minutes, and then bursts out. The parent looks pretty tired. But it recovers quickly and gives birth again in a couple of hours. A healthy *Tokophrya* can perform this miracle as many as twelve times in twenty-four hours. Its name means, in Greek, "the well of birth."

Each newborn *Tokophrya* swims away. Within a few hours it metamorphoses into a young adult, growing a sturdy holdfast, with which it grips the floor of the test tube or the petri dish. It stays put in that one spot for the rest of its life, giving birth to more *Tokophrya.*

This vaguely mammalian style of labor and delivery fascinated Rudzinska. When a paramecium or an amoeba is ready to reproduce, it just splits in two. After each of the halves is full-grown, those split also. Biologists thought of cells like that as virtually immortal. There is never a moment when you can say that one has died. It just goes on and on.

By contrast, *Tokophrya* endures the labor of birth, like us. And day by day, while standing on its holdfast and trawling with its tentacles for food, *Tokophrya* grows old—just like us. So *Tokophrya* makes a good subject for a study of mortality, and because of its holdfast, it is a convenient one; unlike the amoeba or the paramecium, or the yeast cells that swirl around in a pint of beer, each *Tokophrya* stays put. Each mortal poses for the camera all its life. Through the microscope, Rudzinska could watch a single cell on its ride from birth to old age and death and try to figure out what goes wrong inside the cell. It was as if she had the whole problem of life and death on the head of a pin.

When Rudzinska came to New York as an émigré after the war, she worked on other things, including the longevity of the amoeba. She was one of the first biologists to use the new high-powered electron microscope to study the intricate machinery inside cells, those tiny bubbles that keep themselves alive and intact so much longer than a mere bubble of water drifting on a pond. Her microscope back in Poland could make things look one hundred, two hundred, five hundred times larger than life. The electron microscope made them more than fifty thousand times larger than life.

That was useful research, she told me, but as a scientist her heart still belonged to *Tokophrya* and the way it seemed to diagram the mystery of aging. She was convinced that *Tokophrya* would be ideal for the study of length of days. Only a few people in the world were working with *Tokophrya*. She'd lost her own stocks of the creature as an exile and émigré; the stocks that a biologist in Brooklyn shared with her were not ideal for her purposes. She must have told me why, but I've forgotten. What I remember is the tale of her search. She looked everywhere, in ponds, lakes, ditches, and puddles in several states, but she could not find *Tokophrya*, and she could not quite find her way back into the thrill and romance of scientific research that she had felt in Poland during the war. Eventually she fell sick with a high fever, and spent weeks lying in a bed in Rockefeller's research hospital. She thought, *Can this be where my story ends?*

Then, one day in the hospital, while looking out her window, she thought of the fountain pool near her laboratory in Theobald Smith Hall. The pool was not far from York Avenue, but like the rest of the Rockefeller campus it seemed a world apart. It was surrounded by slate walks and marble love seats and ivy-covered sycamores.

From her hospital bed, she asked one of her young laboratory assistants to go down to the pool and collect some water there. She told her assistant to take the water and put a drop under the microscope. And there at last, just as she had hoped, was *Tokophrya*.

That was how Maria Rudzinska recovered her life's work.

After our lunch, she led me to her laboratory. When successful scientists are in the prime of their careers they can command whole floors of prime research space, as Alex Carrel did in his heyday in Founder's Hall. But when they are old and retired, they have to make room for the next generation, Rudzinska explained, ruefully. She worked in the basement of Theobald Smith Hall. Rockefeller's buildings are linked by a system of underground tunnels, and because it was a cold winter day, she led me from the cafeteria through a few of these twisting tunnels until we came to her small, windowless laboratory. She walked slowly and it took us a while to get there.

In her recent experiment, she'd been assisted by two younger scientists, although she usually preferred to work and publish alone. She'd invited both of them to be there when I arrived. They were older than I was, but they looked very young when they stood next to her. They each wore an extra-bright, extra-wide smile that was somewhere between the beaming of reverence for the old master and the beaming of indulgence for the ancient.

Rudzinska explained her latest experiment. With the help of her two young collaborators, she'd collected a few more jars of water from the fountain pool. In her laboratory, they had grown—or, in the jargon of cell biologists, they had isolated and cultured—more *Tokophrya infusionum* in screw-capped tubes. Through the microscope they could see a field of *Tokophrya* waving their tentacles in the water, each one standing on its holdfast. The researchers would search for a single healthy specimen. Using a very fine platinum wire with a tiny loop at the end, like a shepherd's crook, they'd pick out that specimen. They transferred it to a glass slide that had a little shallow depression, called a well. The well was filled with sterilized water.

At the end of each day of the experiment, Rudzinska had inspected the *Tokophrya* in the well through a microscope. It gave birth again and again. She counted the new arrivals, removed the parent with the shepherd's crook, and put it in a fresh drop of water. On Monday, Tuesday, and Wednesday the cell gave birth all day long.

But it was growing older: only one birth on Thursday and one on Friday. None on Saturday.

The aging cell's tentacles weakened, too. Rudzinska was feeding the cell with *Tetrahymena*, which is another protozoan that lives in pond scum. (Through the microscope, it looks something like a hairy mango.) She made sure her *Tokophrya* specimen got just enough *Tetrahymena* every day, not too many and not too few, by directing a stream of them with a fine pipette right at the *Tokophrya*. When it had caught enough food, she would remove the *Tokophrya*, with its prey still in its arms, and transfer it to a fresh well, filled with two drops of sterile pool water.

The healthy cell's cytoplasm was bright and clear. The old cell was dark and shabby, full of dirt and age spots. One of the most famous researchers at Rockefeller, Christian de Duve, had won the Nobel Prize in 1974 for his discovery of a key way the cell cleans up its garbage. He spotted a sort of floating garbage disposal inside the cell, which he called the lysosome, from the Greek: literally "splitting body." Lysosomes swallow cellular trash and digest it. That is one of the cell's great secrets of rejuvenation. The process by which the cell consumes itself in the lysosome is known as autophagy, which means, literally, "self-eating." It is as vital to the life of the cell as eating; but in Rudzinska's aging *Tokophrya* cell, the garbage-disposal system seemed to be failing, too. Everything was failing. When *Tetrahymena* brushed against the elderly cell, they just pulled away and swam on. The cell got darker and weaker, and on Sunday it died.

Rudzinska was looking at a relatively simple life through the transparent walls of its body using one of the most powerful microscopes in the world, and she still could not figure out what was going wrong inside it. The cell died and she did not know why. There were so many possible explanations. At about that time a biologist tried to list them all and counted three hundred theories of aging: genetic theories, evolutionary theories, mathematical and physico-mathematical models of aging. Which one was right? This is what Rudzinska was trying to understand, quietly and patiently. Where was the fatal damage? Was it in the clear jelly of the cytoplasm or inside the dark coiled ball of the nucleus? And why did it happen? Did it have to happen at all?

In retrospect, I was lucky to have met a gerontologist in 1984. I was just in time to catch a glimpse of the slightly depressive backwater that the field had been for generations. "Research on aging, like its subject matter, does not move very fast," as the British immunologist Peter Medawar put it in 1981, when that science was still in the doldrums. "In almost any other important biological field than that of senescence," wrote Alex Comfort, another British gerontologist, in 1979, "it is possible to present the main theories historically and to show a steady progression from a large number of speculative ideas to one or two highly probable, main hypotheses. In the case of senescence this cannot be profitably done."

Comfort was a familiar name to me. Yes, Rudzinska said, Alex Comfort was not only a gerontologist; he was also the author of the worldwide bestseller *The Joy of Sex*.

"We were so angry with him for writing that," she added.

Such a hard and unappetizing problem, aging. The aging of a

living thing is not like the aging of a fine cheese or a fine wine. There the chemistry alters, the molecules change around, and the cheese and wine improve. Nor is aging like the deterioration of a car or a can-opener or any other manmade machine. When gadgets break down they can't fix themselves, and neither can they make more of themselves—whereas a living body, even a microscopic bubble of life like *Tokophrya*, can accomplish both those miracles as long as it lives. Even when *Tokophrya* is ancient, too frail to reproduce, it is still repairing and remaking its own working parts, which is also a kind of reproduction, in a sense; the cell performs the hard work of passing its own body along from one moment to the next, creative work that never stops till death.

So what is aging? Why does the cell stop repairing itself? This is the question that Bacon was asking at the start of the scientific adventure. He knew nothing about single cells but he understood this question. Again, we are so very good at growing and staying in shape when we are young. The mortal body of that single coddled *Tokophrya* would have a chance to last and last if it could only keep up the repairs on Friday the way it did on Monday, when it was young.

Through the microscope, Rudzinska could see so many signs of trouble. *Tokophrya* wears a few coats, or membranes, one on top of the other. Its outermost membrane, called the pellicle, is made of two separate layers that are linked by fine mortise-and-tenon joints. Those joints were popping loose, and the layers were separating. The cell was literally coming apart at the seams.

Of the many studies of aging that she had on her mind in 1984, the most interesting was already fifty years old. In 1934, a biologist at Cornell University named Clive McCay had reported a remarkable breakthrough with laboratory rats. McCay found that if he fed the rats all the nutrients they needed but cut their daily allowance of calories in half, the rats would live about twice as long. Since that time, McCay's discovery had survived test after test. Back when I visited Rudzinska, experimenters were still raising thousands of rats and mice on calorie-restriction diets. The rats and mice got thin and scrawny, but they did live a long time. Nobody knew why.

So Rudzinska investigated the clue of caloric restriction with her *Tokophrya*. Was there something about the reduction of calories that slowed down the metabolic rates of the cells of those mice? Did slowing down their metabolisms make them live longer? She found that when she kept the cells chilly and half-starved, they did live longer.

Rudzinska tried that experiment again and again. She put a single *Tokophrya* in a hanging drop of water on the glass lid of a chamber. Then she fed it, say, three *Tetrahymena*. The next day when she checked on it, it was still healthy and it had produced about that same number of babies. But if she gave a *Tokophrya* forty *Tetrahymena*, it would produce only one baby. If she gave it a hundred *Tetrahymena*, swamped it with fish food, the *Tokophrya* just ate and ate, gorged without stopping. It ballooned out into a giant—dark, opaque, with short, stunted tentacles. It stopped giving birth. It lost its tentacles. And after a few hours, the cell fell apart—cut short in its prime. On the other hand, if she kept a *Tokophrya* on a restricted diet, half-starved for *Tetrahymena*, fed them only one day every two weeks, her *Tokophrya* would live about twice as long.

So calorie restriction worked for species as far apart on the tree of life as mice and *Tokophrya*, which seemed to argue that it might also work for us.

Rereading her papers now, I can see that for all her pains she was a bit isolated, cut off from the news. Most of the names she cites in her papers were already half-forgotten then, biologists who had studied aging in the paramecium and in the amoeba when she was a young scientist in Poland. All around her, biologists at Rockefeller were helping to establish molecular reality; but she did not work with genes and molecules. What you can see inside a cell at 500 or even 100,000 times life size is still coarse compared to what you can see if you get down to the molecular level. The brave new world of molecules was passing her by. And of course she could not begin to explain why multicellular animals like us age, and why unicellular animals seem to escape from aging, or why some multicellular animals do not seem to age at all, like the hydra; while some unicellular animals do age, like *Tokophrya*, even though these two creatures have such a strong family resemblance in body plan and lifestyle that each is like a crude sketch of the other. This kind of confusion is discouraging to scientists, to people who like to figure things out.

Down in the basement of Theobald Smith Hall, Rudzinska and her two young assistants had set up a little demonstration for me. I looked through a microscope on the laboratory bench and saw a whole field of *Tokophrya* standing close together, swaying gently on their holdfasts like a field of alien corn. I turned the knob of the microscope slowly and surveyed the field. There were hundreds and hundreds of cells. It was something to see them, after hearing so much about them, and I looked up to thank the old biologist. Then I put my eye back to the lens. Just when I was about to take my eye away for the last time, I spotted a cell that was shaking back and forth on its holdfast. There was a baby trembling inside it. After a moment, the baby popped out and swam away.

I left the basement laboratory and swung out of Theobald Smith Hall into the pale winter day. Before I walked down the main path to the stone gate on York Avenue, I made a detour through the campus and found Rudzinska's fountain. It had been drained for the winter. A few wet dark leaves from the ivy and the sycamores, the last of the wreckage of the year before, lay plastered to the concrete basin like tea leaves in the bottom of a cup. Somewhere in there, *Tokophrya* lay dormant and encysted, waiting out the winter.

At the time, I found it romantic that science could not answer these elemental and universal questions, questions that must have struck every thoughtful mortal again and again from more or less the beginning of their lives and from more or less the beginning of time. How did we come to be mortal? Do we have to be mortal? What can the science of life do about our mortality? What *is* aging? The image of that microscopic birth in the laboratory still floated before my eyes. I felt as if I had just been granted a glimpse into the fundamentals of birth and death—as if I'd seen as much as anybody could see, looked down to the bottom of the well. No one understood the problem of mortality.

It was clear that Maria Rudzinska loved her work. She loved the questions. She used to sign her letters *Doctor Tokophrya*. But it was also clear to me that she would not be the one to find the answers. And I found that beautiful. I loved the mystery—or else

I'd persuaded myself to love it. Everyone knows that we have to grow old and die, just as surely as everyone hopes for long life. If you drink your cup to the bottom, you reach dregs. If you blunder, if you mess up, if you fumble as you reach for the cup, it spills and it shatters. That is our portion on this planet. The lines of an inscription in Osmington Church, Dorset, carved in 1609, take the shape of the cup:

Man is a Glas: Life is
a water that's weakly
walled about: sinne bring
es death: death breakes
the Glass: so runnes
the water out
finis.

Finis! End of all mortal explanations—whether you think of the problem as spiritual or physical, sacred or secular. We are glass, and we break. We are water, and we spill. We are dust, and to dust we shall return.

That was the problem of mortality as I'd grown up with it. That was the problem of aging with which my generation came of age. Rockets might take us to Mars someday, or out beyond the asteroid belt, but wherever we baby boomers went we would go on bearing the same mortal weight. Rockets might take us to the stars, but only myths could take us to Mount Olympus. We were mortals—and yet the *Eagle* had landed on the Moon.

So we believed in limits, and we didn't—just like the readers of

Mandeville's travels when he described a wonderful bird the size of an eagle in the Egyptian city of Heliopolis, the City of the Sun. The bird is called the Phoenix. "And he hath a crest of feathers upon his head more great than the peacock hath," and his neck is iridescent like "a stone well shining." And the Phoenix lives forever.

Chapter 4

INTO THE NEST OF THE PHOENIX

In ancient legend, the Phoenix was a solitary bird—beyond solitary: unique, one of a kind—that burned itself up in its nest and was reborn. In Egyptian hieroglyphics, the Phoenix represents the sun; in Christian symbolism, the Resurrection of Christ. In Jewish legends, the Phoenix represents the eternal rewards of humility. According to one Jewish legend, Eve offers the forbidden fruit not only to Adam but to every creature in the Garden—the cattle, the deer, the birds. All of them partake except the Phoenix. Only the Phoenix refuses the sin of pride, and that is why the Phoenix is the one creature on Earth that is still immortal. According to another Jewish legend, the Phoenix is made immortal not in Paradise but later, many years after the Fall, for good behavior on Noah's Ark. Noah finds the bird sleeping in a corner with its head tucked beneath a wing. "Why didn't you ask for food?" he cries. The Phoenix says, "I saw you were busy. I didn't want to bother you." And Noah blesses the bird. "Since you were so concerned about my troubles when I was feeding the lions, and when I was trying to figure out what to feed the chameleons, may it be God's will that you never die." From

England to Russia and from Egypt to India and China, people told stories about the Phoenix, which lives a thousand years and then goes up in flames and is reborn to live another thousand, and so on and on forever.

Each of us is that Phoenix. Each of us is one of a kind, and each of us is burned and consumed and constantly renewed and restored. Cells have never come together in the same way to build a body precisely the same as yours; nerves have never met to build a brain the same as yours; the memories that make you what you are have never formed anywhere on Earth or space, in any human skull but your own; and yet all those cells and tender filaments of nerve on nerve are forever falling apart and rebuilding and repairing themselves during every sleeping and waking moment of your life. It's almost as if each instant is our last and first. We are always dying, and always reborn. And that is living. Our bodies are not finished products but works in progress, works continually being dismantled and repaired, rebuilt and restored, destroyed and healed at every moment in the act of living. *Metabolism* is both the building up and the tearing down of the body. *Anabolism* is the constructive part of metabolism; in the process of anabolism we build all of the molecular machinery that we call a living body. Flex a muscle and you encourage the body to build more muscle fibers at that spot. That's anabolism, which can be encouraged artificially with anabolic steroids. *Catabolism* is the destructive part of metabolism, the tearing apart, from the Greek for "throwing down." This throwing down and tearing apart is as much a part of life as the building up. If the Phoenix of the body never did anything but build, it would lose all shape and form; if it did nothing but tear down, it would soon reduce itself to ashes and dust.

If we could only perform this supreme balancing act of death and restoration every day as well as we had done it the day before, tomorrow and tomorrow as well as last year and the year before, then we would be practically immortal. But, alas, with each passing year we perform the miraculous act of the Phoenix less and less well, until at last we die.

The amount of action concealed in that simple word "living" is unimaginable. One single human body is a cooperative of one or two hundred trillion living cells. We have red blood cells that are built to catch oxygen, and white blood cells that are built to catch germs. Rod cells in the eye, built to catch light; hairlike cells in the ear, built to catch sound. Skin cells designed for the palms of our hands, and skin cells designed for the lining of our guts; and stem cells that lie buried in crypts just below each surface, designed to make more of each, each kind of cell patiently replaced, skin for the hands and skin for the guts. Every one of those cells contains thousands upon thousands of working parts: peroxisomes and ribosomes, centrosomes and centrioles, proteasomes and lysosomes, all of them wrapped in membranes within membranes within membranes, and all of them alive. And each of those working parts is made of enormous numbers of molecules, all of them in action like workers at a construction site, day and night.

To understand what's going on in aging, you have to be able to go deep—you have to look into the nest of the Phoenix and into the workings of the cells to see what's going on in there as they build and destroy themselves from moment to moment. That's part of the reason why the science of aging revived in the last years of the twentieth century. At last, decades after Crick and Watson put together their first scale model of the double helix, biologists had the tools to look inside living things at the finest possible level, the level at which all that machinery actually works.

To power all of its molecular machinery, for instance, each cell contains anywhere from a few hundred to a few thousand mitochondria. And every one of those mitochondria contains a large collection of rotary motors. With every breath you take, you set off a long series of actions and chemical reactions that make those rotary motors spin around and around in every living cell of your body like zillions of turbines, windmill vanes, or airplane propellers. These rotary motors turn out a concentrated energy food, an energy-rich molecule called adenosine triphosphate, or ATP. And this ATP, more than any other molecule in the cellular inventory, makes all the rest of the machines go. This is the fuel of all our mortal engines. Without ATP it would be useless for us to breathe in air, to drink and eat. Without ATP, even the smallest piece of action in our bodies would slow down and stop.

Because the mitochondria make ATP, ingredients for this energy food have to be shipped into them. They pass into each mitochondrion through tiny apertures in its membrane. Each of the apertures is equipped with a gate on molecular hinges. The raw materials are shipped in through the gate, and then the ATP is shipped out through the same gate, which swings open and shut day and night. Two Swedish molecular biologists, Susanna Törnroth-Horsefield and Richard Neutze, have spent years studying the mechanics of the hinges of this particular gate. It's not unusual in biology to be that specialized. The living cell is so complicated that there are specialists at every gate. And Törnroth-Horsefield and Neutze can claim with justice that the precise mechanism of their gates' action matters more than most. Their gates are to the life of the body as ports are to a nation. Through these tiny points on the map of each cell, vast quantities of supplies must funnel as they make their way to and from the interior. Most of our metabolites—the raw ingredients of metabolism, and the by-products of metabolism—have to pass through those gates. Although a camel cannot pass through the eye of a needle, write Törnroth-Horsefield and Neutze, it is amazing to think that every single day the camel's weight in metabolites has to tunnel back and forth through a hole that is about a million times smaller in diameter than the camel itself.

Take a breath. As you draw oxygen into your lungs, your red blood cells carry it, molecule by molecule, to every one of your hundred trillion or two hundred trillion cells; and each of those cells transports it down many paths and lanes and through many hundreds of gates and at last through that Camel's Gate, and down into the tiny sealed factory of a single mitochondrion. There the mitochondrion uses the oxygen to produce your energy. Strangely enough, these hardworking mitochondria are the descendants of parasites. They began as bacteria. The bacteria invaded cells that were much bigger than they were, about a billion years ago. Either they invaded, or they got swallowed. Then they made themselves at home in those big cells, and never left. We descend from those big cells with the small bacteria inside them. We are like a people of mixed-race ancestry: animal and bacterial mixed inextricably together. Even now, a billion years later, the mitochondria in our cells still carry their own loops of DNA and they speak their own dialect of the genetic code. In a sense, the mitochondria are still strangers in a strange land, just as they were when they first got lost inside the distant ancestors of our cells. Their alien genes give them the necessary gift of using oxygen in the manufacture of ATP. Those alien genes also encode plans for the tiny rotary motors, which motors revolve at high speeds and turn out the high-energy ATP that they export to the rest of the cell, day and night.

In my first biology class, back in junior high school, I used to try to imagine the oxygen in my breath traveling down into the lungs and the alveoli in the lungs and from there through all the branching capillaries of the arteries until molecules of oxygen reach every single cell. There's so much more to learn about those pathways now. Now molecular biologists have traced what Francis Bacon called "the secrecies of the passages" in almost infinitely finer detail, down through the membranes of the cell and into the mitochondria. In a way, it is sad how esoteric and arcane all this is, our anatomy at the finest level. John Donne when he lay on his sickbed was told by his doctors that he was sick because of vapors. He couldn't see those vapors—he had to take them on faith. Maybe he would die because of vapors. "But what have I done, either to breed, or to breathe these vapors?" he asks, pathetically. As far as he knows, he'd never done anything to go toward a vapor or to draw a vapor toward him, "yet must suffer in it, die by it." The classic lament of the patient whose life depends on doctors' esoterica. Now our fates as mortals rest collectively on studies of molecules that are as alien to most of us as vapors. Francis Crick once said that a good scientist should be able to explain any laboratory result to a barmaid. That's true. But there's so much detail to understand about these molecular machines since Crick and Watson that scientists have trouble even explaining them to each other.

We see so little of the action. We can feel our lungs expand when we breathe in. We can hear our stomachs growl when we're hungry. We can feel our hearts beating. And a few other organs make themselves known to us. But each of these organs has organs. Every single cell is a city. And it is often on the scale of the cell that the real give-and-take of mortal life goes on. That is where the business is transacted. That's where the wheels grind.

The workmanship of all of these miniature machines is magnificent—but they are not quite perfect. Now and then, instead of getting shunted into the rotary motors and turned into useful ATP, a few oxygen molecules fly off like sparks. Almost instantly those oxygen molecules morph into what are called oxidants, or free radicals. Radicals in chemistry are molecules that can swiftly latch on to others; free radicals are loose and wandering and ready to bond wherever they strike. As they wander through the mitochondrion, oxidants damage its working parts in the same way that oxygen in the air will rust iron nails or bring a patina of green to bronze and copper. And this damage accumulates. Because we rust inside, some of our mitochondrial factories break down and stop. Free radical damage to our DNA can cause cancer. In our joints, it can cause arthritis. In the nerve cells of our brains, it may cause Alzheimer's.

So the paradox of mortality is there in every breath we take. We get energy by inhaling oxygen; and we lose energy, breath by breath, day by day, year by year, because of that same oxygen. This is one of the ironies of aging. Oxygen fuels us and oxygen burns us. It is oxygen that makes us go, and it is the very same oxygen that makes us come at last to a stop. Oxygen is double-edged, like the flaming sword that God's angel brandished at Adam and Eve after their expulsion, the sword that turned each way, "to keep the way of the tree of life."

Gerontologists call this the free radical theory of aging. It is a universal theory in that it applies to the deterioration not only of our bodies but the bodies of worms, flies, and every other living thing. The theory was first proposed by a chemist, Denham Harman, in 1956. According to present theory, this is one of the main reasons that our bodies slow and break down with age. Oxidants are perpetually flying out of the molecular works. The machinery they damage most heavily is the gadgetry inside the factory, the mitochondrion itself. So the mitochondria wear out. Their life spans are much shorter than the rest of our bodies. Most of the mitochondria in our cells die and are replaced within less than a month, even the mitochondria inside the cells of the heart, and inside the neurons of the brain, which have to last a human lifetime.

Sometimes oxidants fly into the DNA of the mitochondria. Then they damage its genes. The DNA of mitochondria suffers a much higher mutation rate than the DNA of the rest of the cell, which is ensconced far away from the factory, behind heavy fortress-like nuclear walls. The outer membranes of the mitochondria also get damaged and corrupted. There is much wear and tear at the camel's gates, the gates through which all of those oxygen molecules go in and all of the ATP comes out.

Mitochondria that have been damaged are not allowed to just sit there rusting away in the cell like an abandoned ironworks. Damaged mitochondria are swallowed by machines called autophagosomes, or self-eating bodies, which roam throughout the cell day and night and engulf whatever needs to be disposed of. These autophagosomes swallow so much that they swell like balloons—through the electron microscope their sides look as round and smooth as sausage casings. Then they haul their loads off to the scrap heap: they carry each damaged mitochondrion to some of the cell's giant disposal centers, the lysosomes. A lysosome can dismantle a whole mitochondrion, tearing it to bits in that humble but vital process of autophagy, self-digestion. Lysosomes cut up the ruined mitochondria to be recycled for spare parts.

Gradually our mitochondria wear down more and more, and the body has less and less energy. The rotary motors work less well, all of the machinery in the cell works less well, mistake piles on mistake, and finally we die; all because of free radicals flying like sparks through the mitochondria. Gerontologists call this the mitochondrial free radical theory of aging, or the oxidative stress hypothesis. It was proposed in 1977 by Denham Harman, the chemist, as a refinement of his original theory.

Today most gerontologists agree that this process contributes to our bodies' decline and fall. Every day, you burn through your body's weight in ATP. And every day you manufacture your body's weight in fresh ATP. This is an astonishing statistic. If your body weighs two hundred pounds, you will burn two hundred pounds of ATP today, and you will assemble another two hundred pounds of the stuff to burn tomorrow. A single-celled animal like *Toko-*

phrya will do the same thing on about one hundred trillionth the scale. So will a camel, and so will a blue whale. You are constantly and tirelessly tearing apart not only old mitochondria but every bit of the machinery in the body, all of those gates and hinges and windmills and sluices, every one of your gears and shafts and train tracks and repair robots. And you are rebuilding them just as constantly and tirelessly, night and day. And because you are making little mistakes now and then in tearing down and rebuilding the old factories—the mitochondria—the mitochondria are working a little less well at supplying you with energy, and you are beginning to feel a little tired.

When he began studying aging, informally, in the libraries of Cambridge, Aubrey was fascinated by the invisible internal engineering of the mitochondrion. Thinking about it led him to his first original idea about aging. Very often what wrecks the cell is its failure to recycle its dead or failing mitochondria. Aubrey wondered why those roving disposal systems, the autophagosomes, don't keep up their jobs and dispose of the rusting factories. Why does the cell start out so good at this recycling project and then get so bad at it? Why this slow decline? (The problem of aging in a nutshell.)

It occurred to Aubrey that a roving autophagosome would be most likely to single out a mitochondrion for destruction if its outer membrane were damaged. In fact, damage by free radicals was probably the very thing that marked the aging factory for demolition. But what if a mitochondrion had suffered a mutation that prevented it from making ATP? Then its outer walls would no longer be rusting. A cell would not recognize such a mutant mitochondrion as part of its Rust Belt and cart it away for recycling. The mutant factory would look clean and new from the outside, and it would still be busy on the inside, but it would be useless.

The death of a mitochondrion in a cell might be something like the death of an ant in an anthill. Being an engineer, not a lover of natural history, Aubrey didn't put it to himself this way. But the problem is one that would be familiar to an entomologist. If an ant dies, specialized ants that patrol the tunnels will pick up the corpse and dispose of it. They find the corpse by its odor. If a biologist paints that chemical odorant on the back of a living ant with the tip of a camel's hair brush, then that unfortunate ant's comrades will pick it up and carry it out and dispose of it, alive and kicking. But if the ants were ever to find an odorless corpse, they would ignore it and crawl right by.

Inside a living cell, in Aubrey's hypothesis, the autophagosomes play the part of those disposer ants. If any given mitochondrion accidentally stops the manufacturing process that should have stained it with the telltale mark of age, then that mitochondrion will be undisturbed and unmolested. The power plant really is broken, and it should be scrapped, but it is not scrapped. So that defective mitochondrion multiplies within the cell. Its descendants are also defective but they, too, lack the telltale mark of age. So they make more of themselves, too. Gradually the cell becomes contaminated by all these defective mitochondria, like an anthill filling up with dead and putrefying ants. The cell is sick and oozing with poisons.

Aubrey laid out this argument, which had many twists and

turns, in two technical papers in the late 1990s. It was an interesting but unpleasantly complicated hypothesis. It was ugly, as he himself was the first to admit. If true, it would resolve a few problems with the existing theory, and maybe introduce a few new ones. After publishing these papers, Aubrey wrote a technical book about the whole subject, *The Mitochondrial Free Radical Theory of Aging*. For this work the University of Cambridge awarded him a Ph.D. in biology in the year 2000. (The university gives Ph.D.s to its graduates if they do suitable work on their own.) Aubrey was now an authority on aging mitochondria.

Along the way he met new heroes and kindred spirits.

One of them was Denham Harman, born in 1916 and still at work when Aubrey entered the field of gerontology toward the end of the century. (Harman is still going strong as I write, in the fall of 2009, at the age of ninety-three.) For decades, Harman had been trying to persuade more scientists to explore the causes of aging. In 1970 he founded a new association, the American Aging Association, which goes by the acronym AGE. It is a society of scientists who focus on the study of aging and its cure, and publish the peer-reviewed journal *Age*. Aubrey felt stirred and inspired when he listened to Harman, or to Richard Miller, another leader of the field, who opened one international gerontological conference with the simple words, "Aging is bad for you." Optimists like Harman and Miller surveyed the spectacle of the body's almost endless self-consumption and renewal and felt hopeful. Aubrey read Harman's studies of the potential of antioxidants: natural or artificial compounds that can soak up the free radicals in the body and prevent their doing so much damage. He read Miller's studies of calorie restriction. Somehow the two were clearly connected. If we eat less we burn less; the metabolic fires slow down and there are fewer free radicals shooting around in the cells like sparks.

After all, we are looking at a body that is constantly and creatively and minutely maintaining itself as long as the body is alive. Even our skeletons are alive. Every one of our 206 bones is constantly being torn down, restored and remodeled like every other part of our living machinery. Cells called osteoclasts carve away the old bone, and cells called osteoblasts build up the new. As at most construction sites, the demolition goes faster than the rebuilding. Three weeks per site for demolition; three months for reconstruction. Even in our bones, a careful balance of creation and destruction, day and night. Too much destruction and we develop diseases like periodontitis, rheumatoid arthritis, osteoporosis. Too much creation and we suffer, too, as in the rare condition osteopetrosis, literally bones of stone, in which the osteoclasts fail to do their jobs and the osteoblasts make the skeleton increasingly dense, heavy, and brittle; or another rare condition, fibrodysplasia ossificans progressiva, in which even muscles, tendons, and ligaments turn to bone.

Our very skeletons are full of youth, and fight all the way down to the dust. Like the Phoenix, we destroy ourselves and restore ourselves—burn ourselves down and build ourselves up—not every thousand years but daily and hourly—all the way down to the bone. Life seen this close up looks like a kind of bonfire, like the flames of the Phoenix when it self-immolates in its nest. But the Phoenix of legend is immortal and we are not. Why did life evolve this way, so that the miracle of the resurrection succeeds brilliantly in our youth and then fails? Why not perfect renewal of the body forever?

Life is a kind of sacrifice, a sacrifice we have made from the beginning, and make every day of our lives. Each mortal body is a story of sacrifice and renewal that slowly fails. It is a very old story for which we all, at least at one time or other, would love to change the ending.

PART II

THE HYDRA

They are ill discoverers that think there is no land, when they can see nothing but sea.

—FRANCIS BACON,

THE ADVANCEMENT OF LEARNING

Chapter 5

THE EVOLUTION OF AGING

Not long ago I was talking about the problem of mortality with a physicist and he told me, with a smile, that it's in the nature of everything to fall apart. That is what the law of entropy tells us about inanimate objects like planes, trains, and automobiles. That's also what common sense tells us about animate objects like our own warm, breathing bodies. But we don't fall apart between the years of, say, six and twelve. We grow bigger and stronger in those years. If we can do that much when we are growing, then why can't we at least hold steady, hold our ground, from the ages of twenty to a hundred and twenty? We don't, but that doesn't mean our failure to do so is mandated by the laws of physics. If that's breaking the laws of physics and common sense, then we've already broken them. Every human body breaks those laws in the womb from the moment sperm meets egg. Those two microscopic cells meet in the dark and nine months later, after a miraculous construction project, a baby is born with a body made of trillions upon trillions of cells, from the brain cells inside the still-soft skull to the skin cells in the ten fingers and ten toes. And the history of the development of life on

Earth is at least as spectacular as the development of each life in the womb. Life on Earth, from small beginnings, has attained extraordinary profusion. Three billion years ago, life was all microscopic single cells. And now there are millions of species of living things, from shrimp to whales, from mites to elephants. The development of life on Earth is like the development of a life in the womb: it defies common sense, and the intuitions of physicists, like a ball that rolls uphill.

If life can do so much in the first half, why does it fail in the second? Why can't it keep the ball rolling? Bacon makes this point in the first pages of his *History of Life and Death.* He chastises the physicians and philosophers of his time for missing it. Conventional wisdom in Bacon's day held that there is something in the body that can't be repaired, some "radical moisture" that can never be replenished. Our bodies lose that moisture and dry out and that's why we get old. But that idea is "both ignorant and vain," Bacon writes; "for all things in living creatures are in their youth repaired entirely; nay, they are for a time increased in quantity, bettered in quality." So much so that "the matter of reparation might be eternal, if the manner of reparation did not fail."

We grow up, and then we seem to hold steady for years. A woman between the end of puberty and the onset of menopause balances the building up and the tearing down of her bones so perfectly that they grow neither too heavy nor too light. Her whole body—flesh, blood, bone, and sinew—is a kind of fountain in which the new continually replaces the old and the form stands as if it would stand forever. Then, after menopause, the balance fails, and bone mass declines, and osteoporosis sets in. But why does the balance have to fail? Why did this failure evolve? Which is to ask the most fundamental question of the science of mortality: How did old age and death come into the world?

The answer that has now emerged in the science of mortality gives hope to the field's optimists.

Darwin himself does not seem to have thought about this question. Apparently, like most of us, he took aging for granted. But one of Darwin's first great supporters, the German biologist August Weismann, did think about it. His own conclusion was dark—so dark that it may have contributed to the doldrums that gripped the field for much of the twentieth century.

Weismann laid out his argument about the evolution of aging in one of his first lectures as prorector of the University of Freiburg in the spring of 1883. He published the lecture that summer as an essay, "Upon the Eternal Duration of Life." "In my opinion," Weismann said, "life became limited in its duration, not because it was contrary to its very nature to be unlimited, but because an unlimited persistence of the individual would be a luxury without a purpose." In other words, he believed that life on Earth had been immortal, once upon a time. Immortality was just as natural a state for living creatures as mortality. "Among unicellular organisms natural death was impossible," Weismann wrote. An amoeba and the paramecium never die only because they can't—because they are too simple to die. But as soon as multicellular life evolved on this earth, aging did become possible for them, and they began to grow old and die.

In the beginning, in Weismann's view, death did not exist; and then life invented it. In fact, if human beings ever did find a way to make ourselves immortal, said Weismann, then our descendants would just evolve mortality all over again. "Let us imagine that one of the higher animals became immortal," Weismann writes; "it then becomes perfectly obvious that it would cease to be of value to the species to which it belonged." Think of it this way, he says. Even if a tree or an elephant or a mouse never got killed by some accident, even if it lived for eternity—which is, of course, impossible—it would be bound to get damaged and then crippled by this and that affliction, somewhere along the line; "and thus the longer the individual lived, the more defective and crippled it would become, and the less perfectly would it fulfill the purpose of its species." The species would have to keep producing new and healthy specimens to take the place of its sick, hobbled, and infirm; "and this necessity would remain even if the individuals possessed the power of living eternally."

So death is a sacrifice that each generation has to make for the sake of the next. We reproduce, and then we have to die. "Worn-out individuals are not only valueless to the species, but they are even harmful," he says, "for they take the place of those which are sound. Hence by the operation of natural selection, the life of our hypothetically immortal individual would be shorted by the amount which was useless to the species." Life's invention of death proved to be so successful and necessary, death made species that possessed it so vital, that once death arose it became universal; so that "the higher organisms, as they are now constructed, contain within themselves the germs of death."

In Weismann's view, then, aging and death are accomplishments that we complicated creatures should be proud of. Amoebae and other single-celled organisms are forced to remain immortal because they do nothing but divide and divide. But the immortality of protozoa is primitive compared with the mortality of metazoa like ourselves.

There's a certain fascination in this idea, dark as it is. In Weismann's view of life, aging is an adaptation. Death itself is an adaptation. Death is more important to us than eyes, ears, teeth, and hands; or flukes, gills, and flippers; or roots, branches, and green leaves. Just as a beetle never grows as large as a horse, because there are natural limits to its growth, so a beetle never lives as long as a horse, and a horse never lives as long as a man, because there are natural limits to their longevity.

Besides the protozoa, Weismann did recognize one other form of biological immortality on this earth. Our bodies are divided into two kinds of seeds, two kinds of cells, the mortal and the immortal. The seeds in our eggs and sperm have been passed down to us from generation to generation. Weismann called these seeds the germ cells, and the rest of our bodies the soma. The soma is doomed, but our germ cells are potentially immortal.

That part of his argument is still regarded as well established. But his basic premise is not, even though most of us still assume that it is true, because it makes intuitive sense. If asked why we grow old and die, most people today would answer, just as Weismann did, that we have to wear out and die to make room for the next generation.

And most biologists in Weismann's generation and for several generations afterward did think his point made sense. Weismann's argument helped inspire Sigmund Freud's famous theory of the death instinct. "What lives," Freud wrote in *Beyond the Pleasure Principle*, "wants to die again. Originating in dust, it wants to be dust again."

The biologist who spotted the flaw in Weismann's argument was Peter Medawar, who won a Nobel Prize for work in immunology during World War Two, when he developed new methods for skin grafts. A few years after the war, Medawar published two celebrated essays on the problem of aging, "Old Age and Natural Death" and "An Unsolved Problem of Biology." There he both posed and solved the problem of aging, in the view of most gerontologists today; he explained why evolution brought old age and natural death into the world. At the time I visited Maria Rudzinska, back in 1984, Medawar was by far the greatest living scientist in their still-small field, even though the problem of aging was only one of a vast number of his interests. That year, he gave a public lecture in New York—at the Explorers Club, I think—and I went to hear him. He was a handsome, elegant, and sophisticated old man, crippled by a stroke. He lectured from a wheelchair, with his equally elegant wife standing at his side.

Medawar had studied Weismann's argument about old age and decided that Weismann was completely wrong. In his essay "Old Age and Natural Death," Medawar quotes those wise-sounding lines of Weismann's about worn-out bodies, which are useless to the species, and even harmful, because they get in the way—so harmful that even if their ancestors had once been immortal, natural selection would have shortened their life spans and made them mortal.

"In this short passage," says Medawar, "Weismann canters twice around the perimeter of a vicious circle. By assuming that the elders of his race are decrepit and worn out, he assumes all but a fraction of what he has set himself to prove." Why *are* they worn out? That's the whole question, says Medawar. That's Weismann's first canter around the vicious circle. And if bodies *are* worn out, then natural selection will weed them out. Bodies don't have to invent or evolve an elaborate adaptation like Death by Old Age to take themselves off the stage. Give mortal bodies enough time on this earth, and sooner or later a cold winter or a hot summer, a drought or a flood, a famine, a pestilence, the wolf at the door, a chicken in the snow, or any one of nature's myriad dangers will come and find them. Plain bad luck will take them out.

Mother Nature is infinitely inventive when it comes to fatal accidents. Because we have managed so successfully at insulating ourselves from most of them, we forget how tough it is out there, even for creatures that are young and healthy. Darwin makes this point in the most famous chapter in the *Origin of Species*, "Struggle for Existence," which begins, "Nothing is easier than to admit in words the truth of the universal struggle for life, or more difficult—at least I have found it so—than constantly to bear this conclusion in mind."

As Darwin goes on to say in "Struggle for Existence," most animals die young. Take wild mice. Nine out of ten wild mice die before they have lived a year. Some of them get pounced on by a cat or an owl. More of them die of the cold, at night, hungry and shivering. They die because they don't have enough fuel in their bodies to keep warm. If you think of the life span of a mouse as having Seven Ages, like a man, then most wild mice don't survive beyond the young lover in *As You Like It*, composing a sonnet to his mistress's eyebrow. They don't die of old age; they die huddling together for warmth in the long hours of the night. Virtually no wild mouse is ever so lucky as to survive to extreme old age, or about three years, which is the age when a well-fed mouse in a safe warm cage finally totters to its end, "sans teeth, sans everything."

It's the same with gray squirrels. They can climb trees to get away from cats and dogs and kids with slingshots. But even so, only about thirty in a hundred survive longer than one year. Only six or seven in a hundred survive more than four years. And yet when gray squirrels are kept in zoos they can sometimes live twenty years.

From the fact of the struggle for existence, Darwin drew a conclusion that seems simple in retrospect. Darwin's process works by selecting slight variations—those that make a difference in the survival of an individual. There are times when the slightest variation will determine who lives and who dies; who gets to reproduce and who dies without passing on the genes.

And from this same hard fact of life, Medawar drew a second conclusion. In the wild, life is so hard that variations are weighed in the balance, with the best selected and the rest rejected, when the individual is *young*. It is only among the young that variations will be weeded out. Those that appear later in the creature's life span won't be culled, because the creature will almost never live that long anyhow. Again, as a general rule, life in the wild is so dangerous that no matter how fit they are, most creatures don't live long enough to grow up, let alone grow old. Most don't live long enough to pass on their genes. "We behold the fact of nature bright with gladness, we often see superabundance of food," Darwin writes in "Struggle for Existence"; "we do not see, or we forget, that the birds which are idly singing round us mostly live on insects or seeds, and are thus constantly destroying life." They are cutting short the lives of many of those bugs and plants before they make more bugs or plants. And if the birds themselves run out of bugs and seeds, they die young, too.

You can convince yourself that most wild things die young by doing a simple thought experiment. Suppose, says Darwin, an oak produced only two seeds a year—"and there is no plant so unproductive as this." If each of those two seedlings grew up the next year and produced just two more, and each of those produced two, and so on, and if each of the seeds germinated, then in twenty years that first oak would have produced a forest of one million oaks. Or take elephants, which are the slowest-breeding animals on the planet. Elephants start breeding at the age of thirty. If one pair began breeding when they were thirty and produced only six baby elephants by the time they were ninety, and if all of those babies grew up and bred, then, well before a thousand years had passed, that matriarch and patriarch would have produced a herd of almost nineteen million elephants. If oaks and elephants went on like that the whole planet would soon be oaks and elephants. What this arithmetic suggests is the brevity of life throughout all its kingdoms. Most animals don't live long enough to become parents. Most seeds don't live long enough to set seed themselves. One spring, Darwin tested that point with an experiment in his garden. He marked out a plot of ground three feet long and two feet wide. He dug and cleared it and counted all the weeds as they came up. Out of 357 weed seedlings, he says, 295 were destroyed, most of them chewed and swallowed by slugs and bugs. Those seedlings never passed on their genes. Even weeds die young—which is why they don't completely take over the planet either.

Since oaks and elephants and dandelions never take over and engulf the earth, Darwin concludes, we may be sure "that this geometrical tendency to increase must be checked by destruction at some period of life." And that period, Medawar adds, is *youth*. Death hovers everywhere and prevents the conquest of the planet by oaks or elephants or anything else alive. Everywhere in the living world a lucky few survive while the rest die young.

So that, Medawar argues, is why animals and plants are mortal: that's why they get frail when they get old. It's not that elephants and dandelions have evolved progressive frailty as an adaptation, to get themselves off the stage of life and make way for new elephants and dandelions. It's not that death is an adaptation. It's just that genes that cause progressive frailty do not matter in the wild. Genes that cause late-onset diseases are invisible in nature. They don't matter because animals and plants almost never live long enough for those problems to develop. Way before they reach the age of late-onset diseases, they are long dead anyway. Think of those mice in the fields and woods. Nine out of ten of them will die before they are one year old. If they put their energy into building bodies that will last longer than a year, only one in ten will profit from the investment, and nine out of ten will be the poorer for it. They'll be less competitive because they wasted resources they could have used when they were young. Why put your precious energy into building with the best materials if your time is so short? The wolf at your door will blow down a house of bricks just as fast as a house of wood or a house of straw. Why plan for a retirement that only one out of ten will live to see? Better use every drop of energy for hustling and making babies. And it is literally energy we're talking about. Our mortal bodies do a huge amount of work maintaining our libraries of DNA, and scraping away the rust to keep ourselves from browning inside like sliced apples, and just keeping warm. All that manufacturing, importing, and exporting of ATP. Most of us have no idea the Herculean effort required by the body to find all that energy and pay for it, to keep up with the heating bills and the repair bills.

This was Medawar's great insight. It's a striking conclusion, almost as broad as the conclusion that Darwin drew when he contemplated the struggle for existence. Like Darwin's argument, it applies everywhere in the tree of life. What's true for elephants and dandelions would have been true for our own ancestors, too, until we invented civilization and saved ourselves from life spans that were nasty, brutish, and short. As Darwin says in the *Origin of Species*, the slightest variations will sometimes determine who shall live and who shall die. Those of our ancestors who could see the lion first and outrun it fastest survived to make it back to the comforts of the cave or the tent, and the arms of their mates that night. That is why they had the chance to become our ancestors. Those who did not live long enough to be parents are not among our ancestors.

And our ancestors' genes are now our genes.

Think of Shakespeare's Seven Ages of Man: the infant, the schoolboy, the young lover, the soldier, the judge, the retiree, and then the senile old man, withering back toward nothingness. Among our ancestors in the wild it was only in our First, Second, or Third Age that crippling variations would be weeded out. Variations mattered only up until the age of the young lover. And that's still true today, on average, even though we no longer live in the wilderness. Suppose, for instance, that you are born with a mutation that will cause trouble only when you reach the Fourth Age of Man. You carry a defective gene that won't do you any harm at first, but later on will make you very sick. You are just fine as a mewling and puking baby, you are healthy as a schoolboy, and as a lover writing sonnets to your mistress's eyebrow. You are still fit and strong at the fourth age when you play the part of the soldier charging the cannon's mouth. But just as you reach middle age, the fifth age, the age of the judge sitting on the bench, you begin to have fits. You thrash your arms around in the air. You attack the plaintiff when he approaches the bench. You go home and attack your children. You start speaking in tongues. You are suffering from Huntington's disease, which is caused by a mutation in a single gene on Chromosome Four. It is a horrible, progressive, fatal disease, and there is no cure. You will not live to your Sixth Age, the age of the retiree with the spectacles on the nose, the age of the shrunk shanks and slippered pantaloons (as Shakespeare pitilessly puts it). But you have already passed on your genes, just as your father or mother passed them on to you. If you have two children, the odds are that one of them has Huntington's disease. Natural selection cannot prevent that Huntington's gene from passing this way through the generations, century after century. Darwin's process could stop the spread of Huntington's disease only if the mutation made people sick in the first half of life, when they are most likely to become the fathers or mothers of children. After that, whatever the mutation does, it goes unpunished.

Fortunately, Huntington's is a rare disease. But the same argument applies to each and every gene that causes human bodies harm when they are beyond the Third Age, the age of the young lover. Suppose you carry genes that maintain your muscles when you are young and then allow them to weaken and shrivel away in your forties, fifties, sixties, and seventies. Genes that allow old muscles to shrink are very common in the human species, and so is the condition. Geriatricians call it sarcopenia. It is one of the most common problems of old age. Or suppose you have genes that allow the lenses of your eyes to stiffen as you reach the age of forty. Genes that fail to prevent that condition are extremely common, and so are reading glasses. Almost all of us carry genes like those and pass them on to our children. We're horrified by a rare disease like Huntington's, but we all carry innumerable genes that let us develop problems around middle age, and in the end they're fatal, too.

In our comfortable civilization, we can put up for a long time with weaker arms and legs and back and weaker eyesight. We can survive with them even into our eighties or beyond with the help of walkers and glasses and cataract operations. Back when we lived in the wild, as our ancestors did for millions of years with bodies very much like those we have now, those conditions would have been just as fatal in middle age as Huntington's is today. Back then you had to run away from a lion. But once again, Darwin's process would have been powerless to weed out those genes, for the same reason that it could not weed out Huntington's. Those are all problems that start to bother us long after our bodies have reached puberty. By the time our muscles and our eyes are weakening badly, by the time our necks and backs begin to bother us, most of us have passed on our genes to our children. Darwin's process, evolution by natural selection, the process that gave us all of this miraculously intricate living machinery, cannot prevent most of that machinery from beginning to slow down and fall apart in the Fifth, Sixth, and Seventh Age of Man.

And of course what is happening at the level we can feel and see and notice, the level that bothers us—the stiff neck, the stiff knees, the dry skin, the brittle fingernails—is happening inside the body, too, where the machinery is far more intricate than the stuff we can see and feel. Darwin's process gave the cell the machinery of the Phoenix, the machinery of repair and self-renewal. It gave us the mitochondria that produce our energy and the autophagosomes that clean up the mess. But Darwin's process cannot prevent that beautifully intricate machinery from slowing down in our forties and breaking down in our eighties. Some of the slow failure of our muscles begins there, in the failure of their mitochondria. But again, by the time sarcopenia starts to bother us, we have long since passed on our genes.

Our bodies are capable of producing a state of extraordinary health and stability. If we could stay at that stage of health, the stage of the Second Age of Man, when we are about twelve, then, according to some actuarial estimates, we would live, on average, for about 1,200 years. One in a thousand of us would live 10,000 years. But in the wild, our distant ancestors could not expect to survive past the age of one or two, and only the very lucky reached the age of twelve or twenty. So our bodies put everything they have into making it to twenty, and the rest be damned.

That is why the Phoenix burns so brightly in our youth and then begins to burn down, like a small flame aglow on its own ashes.

* * *

Medawar thought he was burying Weismann with this argument, but in fact the two biologists' ideas bear a strong family resemblance. Medawar's is a story of sacrifice, too. In Medawar's vision, as in Weismann's, each generation dies for the next. According to Medawar's argument, the only mortal bodies that pass on their genes are those that are quick to reproduce—to get into the game while they are still among the living. In other words, our bodies are built to grow up fast. They aren't built to last.

In some ways the sacrifice in Medawar's story is even more painful. As Medawar pointed out, his argument has an awful wrinkle. Any gene that helps you grow up fast in your teens will be favored by natural selection even if that same gene turns on you and kills you later on. If it helps the body with quick-and-dirty construction in the womb, or during its first twenty years of life, then that gene will be likely to be passed on, even if it makes the shoddy body fall apart in forty years. As long as Jack and Jill can get up the hill, it doesn't make any difference if their genes make them tumble down.

It's bad enough that evolution allows you to pass on a time bomb like Huntington's. Evolution may actually encourage you to pass on such time bombs. Natural selection may favor time bombs. If they help you in some way to reach puberty fast, then they will be favored because you will be more likely to survive long enough to pass them on. By the time you've reached your fifties and those bombs begin to explode, you've long since passed them on to your babies. And again, in the wild, the odds were against reaching your fifties anyhow.

This is a chilling vision when you take it in. It gives new meaning to the expression "over the hill." On tall steep tropical islands that lie in trade winds, the winds almost always blow from one side.

The windward side of the island is often wet with rain because that's where the clouds form, while the leeward side is usually dry and barren because the rains have already fallen by the time the wind reaches that side. Rain falls only on the windward side. The leeward side of the island is stuck in what is known as a rain shadow. So half the island stays wet and green and young, and the other half stays dry, bare, and old. That is how it is with us, if Medawar's argument is correct. As soon as we are just one step past our peak, we begin to descend into the shadow of Darwin's mountain. We descend from the green crest, and we walk down toward the valley of the shadow of death.

Medawar's argument has gotten more and more support since the middle of the twentieth century. In the late 1950s, the American evolutionary biologist George Williams reviewed Medawar's logic and agreed with him that aging is a surprising feature of life, a feature that can't be explained as Weismann did by calling it an adaptation. If an embryo can grow into an adult and an adult can keep itself up for decades, then why can't the adult keep itself up indefinitely? "It is remarkable," Williams wrote, "that after a seemingly miraculous feat of morphogenesis a complex metazoan should be unable to perform the much simpler task of merely maintaining what is already formed." Williams agreed with Medawar that each line of life must carry genes that help it to grow up, and then turn around and betray it and bring it down.

In the late 1970s a British biologist, Tom Kirkwood, put this evolutionary theory of aging in contemporary terms in a fresh series of papers. Kirkwood gave this argument a memorable name: the theory of the disposable soma. Once we are past the age of reproduction, once we are no longer making babies and raising families, our bodies become disposable. Once we've passed on our genes, we're trash.

Gerontologists have recognized this nightmarish possibility in the theory for decades, and they have proposed a few examples, most of which have yet to be proved conclusively. An interesting example has been proposed by Caleb Finch, of the Andrus Gerontology Center at the University of Southern California, Los Angeles. Finch is one of the greatest scholars among gerontologists today. He argues that inflammation may be a crucial problem in aging. Since the year 1800 or so, we have increased the average human life span on this earth by 100 percent. We have reduced childhood mortality by 90 percent. Since 1850, we have also reduced mortality in old age, with most of the gains in the last few decades. Most gerontologists attribute these epic, century-by-century victories to the broad progress of medicine, to the growth of economies, to near-global improvements in nutrition. But Finch and his colleagues argue that what may matter most in this story is quite specific: we catch fewer infections when we are children. Infections can cause chronic inflammation, and Finch believes that inflammation may be the single most important factor in the decline of old age. Chronic inflammation is now thought to increase one's risk of heart attacks, strokes, cancer, and even, possibly, Alzheimer's disease.

As Finch notes, some infections cause long-term damage directly. For instance, childhood strep infections, if untreated, can lead to rheumatic heart disease, and the damage to the heart valves can be fatal decades after the strep. But in millions of cases the links of cause and effect may be more subtle and may show up only in statistics. Cohorts of babies with high levels of infant diarrhea and enteritis, for instance, have been found to have more heart problems and more respiratory problems when they grow up. If you are an American in your fifties, and you had a major illness as a child, you are 15 percent more likely to have a heart condition, and you are twice as likely to have a chronic lung condition. You are also twice as likely to have cancer, although no one knows precisely why. Finch thinks all this damage may be done by elevated serum levels of certain inflammatory proteins, such as C-reactive protein (CRP). People who live in places where they are likely to be exposed to chronic tuberculosis, diarrheas, and malaria are likely to have elevated levels of CRP throughout their lives. This is why your dental hygienist is always reminding you to floss. The inflamed gums of periodontal disease can cause chronic high levels of CRP, and, it's now thought, raise your risk of heart disease, stroke, and cancer.

It may be that the body's response to short-term infections when we are children works to help us recover quickly when we are young, but then the inflammation lingers in ways that make us sick when we are old. If so, that would be an example of the kind of juggling act that the evolutionary biology of aging predicts. If what heals us as babies makes us sick as old bodies, then evolution will favor the healing of the young and ignore the damage to the old.

That may be why the elderly in developed countries started living longer in the last decades of the twentieth century. They may have lived more years in old age because they'd contracted fewer infections when they were very young, in the early decades of the century. They were cleaner, better-fed, and better-doctored as children, and their bodies had lower levels of inflammation for the rest of their lives.

(On the other hand, many people think our health is *suffering* because we are kept too clean while we're young. The rise of asthma in developing countries may be caused by our lack of exposure to pathogens early in life. That's the very opposite of Finch's idea of early exposure and inflammation.)

In any case, if you're a body and you've got to survive long enough to reproduce and you've got limited resources, then you're going to put those resources into the task of reaching reproductive age, finding a mate, and passing on those genes. If you divert too many of your limited resources into building a body that will last into old age, then you may not live long enough to pass on your genes. Bodies that follow that losing strategy will get weeded out by natural selection. In this way, evolution selects for decrepitude. Yeats said that each of us is forced to choose between perfection of the life or perfection of the work. In a sense, our genes choose perfection of the work—the work of reproduction, that is—rather than perfection of the life—the long life well lived. Our genes made this choice in the time of our distant ancestors, long, long ago and far away. Now our bodies make the sacrifice whether we like it or not.

Medawar himself was not the kind of man who found it comfortable or easy to step aside as one stage of life advanced to the next. One of his adages was, "Humility is not a state of mind conducive to the advancement of learning." In a memoir that he wrote in his old age, he confessed that he had put his scientific life far ahead of his family life. "I had thought, when I was a boy, that I should be a good father, one who wisely and kindly guided my children, shaping their minds and morals by imperceptible degrees. My performance fell so far short of these ambitions that I was an outstandingly rotten father and neglected my children disgracefully."

Medawar had also fought retirement. He'd gone back to the lab after a first stroke in 1970. "No working scientist ever thinks of himself as old," Medawar said. He kept working in spite of not one but two grand retirement parties. In an after-dinner speech at the second retirement, Medawar told a long table of colleagues that it was his ambition to stay on until he'd become a notorious pest. "I hope to continue working until as I career down the corridors in my electric wheelchair, newcomers flattening themselves against the wall will say to each other, 'That's Medawar, you know: they simply can't get rid of him.'"

"We're saying that already, Peter!" cried a voice from the other end of the table.

He died a few years later.

Bleak as it is, Medawar's view of life does have some hopeful features for us, because our species may be a special case. Those of our ancestors who survived and stayed fit as grandparents, for instance, might have been able to help their grandchildren enough to make a difference for their survival. If they were wonderful grandparents, then their longevity genes would have been more likely to be passed down. Families fortunate enough to be blessed with those grandparents and longevity genes would have been more likely to grow and thrive, generation after generation. And this process might have been self-reinforcing: not a vicious circle, but a virtuous circle. Because of our big brains and our gift for culture, our Old Ones would have retained a value to their kith and kin that old chimps or apes would not. The value of the older and wiser heads among us would have grown the more human culture grew—that is, the more there was to know. Consequently those of our ancestors with genes to make them last a long time would have tended to pass on those genes through their grandchildren. Consequently we evolved to last longer and longer. This virtuous circle might help to explain the longevity of grandmothers and grandfathers. They are able to help their sons and daughters raise children of their own. This is known as the Grandmother Hypothesis.

To study the evolution of human life spans, some physical anthropologists have made a specialty of Paleolithic dentistry. By inspecting the wear on our ancestors' teeth, they can estimate how long the owners of those teeth used them. They count baby teeth, adult teeth, and examine through microscopes the wear and tear on the molars, because the life of hunters and gatherers puts a lot of stress on chewing. Recently two anthropologists, Rachel Caspari, at the University of Michigan, Ann Arbor, and Sang-Hee Lee, at the University of California, Riverside, examined the whole Paleolithic dental database and made a provocative discovery about the evolution of human life expectancy.

In the past, anthropologists have been hampered in such studies because the number of people who lived to a ripe old age in the Stone Age was very small, and because by the time they reached old age, our ancestors' teeth were in such bad shape that they no longer provided much of a fossil record. There are huge gaps, so to speak, in their fossil dental records. So anthropologists had assumed that they would never be able to use fossil teeth to trace fine-scale trends in the evolution of human life expectancy. They could not do so because the evidence for assessing the maximum life spans of our ancestors was gone. If any of our ancestors survived to the Seventh Age, "sans teeth . . . sans everything," they had no teeth left to bequeath to science. Nothing left for Caspari and Lee to study.

But Caspari and Lee reasoned that to reconstruct life expectancy you don't have to count the Seventh Age, the age of the oldest old. By definition, the oldest old are a tiny minority. They don't matter much to the overall pattern. All you need to know is how many people in the Stone Age lived to be old. So Caspari and Lee reexamined the Paleolithic dental records, focusing on the third molars—commonly called the wisdom teeth—which erupt at about the time that we reach puberty. They studied all of the fossil teeth they could find in the biggest sample ever analyzed of Stone Age skeletons, and they sorted them into three groups: children, who died before their third molars had erupted; young adults, who had those molars but showed very little wear on them; and older adults, defined as people whose molars were worn enough that they had lived at least fifteen years with wisdom teeth. Thirty is a significant year in a life cycle where fertility begins at fifteen. If a girl became enough of a woman to have a baby at fifteen, she would be old enough to become a grandmother at the age of thirty. Boys who fathered children at the age of fifteen or so could be grandfathers at thirty.

Caspari and Lee found that in the Upper Paleolithic, which began roughly thirty thousand years ago, more and more of our ancestors lived to be old. In the course of the Upper Paleolithic, the number of adults who lived long enough to be grandparents rose fourfold.

The very ability to live to older ages may have been fostered by the culture that grew and deepened as increasing numbers of elders helped to raise the children of the tribe and pass on what they had learned. And so the two may have advanced together, the survival of the young and the survival of the old, literally hand in hand.

This advance in human life expectancy may explain why populations in the Upper Paleolithic suddenly grew in numbers and went trekking farther and farther into the landscapes around them. Generation after generation carved new trails, built new settlements. It may be that the elders' investments in the children gave these tribes and villages an advantage that helped them grow and prosper and wander. The change began about 30,000 years ago, and by about 15,000 years ago we had colonized almost all of the planet. Anthropologists have argued for decades about what caused this extraordinary expansion of culture and geography, which they sometimes call the creative explosion. Was our ancestors' increase in longevity, our growing length of days, at the heart of the change? Did our longevity help lead to what we think of as modern human life? We may never know which came first, the length of days or the improvements in human culture and society that led to the length of days. They too may have advanced hand in hand. "We suggest," Caspari and Lee write, "that this increase in longevity addresses the meaning of modernity itself."

If this argument about our evolution is correct, and longevity has played a part in the human success story from the beginning, then we are partial exceptions to the rule of the disposable soma. We are worth something in our old age, after all. Our longevity is of adaptive value. If so, then some of the changes in our aging bodies may be adaptive—including the big one we call The Change. Why is it that women approach the end of their fertility at the age of forty-five or fifty, when they are still relatively healthy and fit? It is true that they are running out of eggs, but why do they have to run out so young? Men still have sperm. The human reproductive system ages faster in women than the body as a whole, and by age forty-five, according to one authority, it "can be said to be in the state that a woman's other organs have reached by eighty."

Menopause may be a by-product of the evolution of our particular niche as a species. The Change may have helped us succeed in our niche as knowledge-gatherers. The argument runs like this. Those among our ancestors who were better knowledge-gatherers survived longer and had more offspring, who survived longer yet. So we evolved bigger brains. Birds and bats have wings, tortoises have shells, we have big brains. We evolved on the African savannah among dozens of sibling species of primates. Our unusually big and agile brains gave us the ability to cope with predators and prey, and to share ridiculous quantities of useful and useless information. And our life spans evolved until they were much longer than any other primates on the planet. But the development of our big brains presented evolution by natural selection with a difficult engineering problem. Babies with bigger brains have more trouble passing through the birth canal, whether headfirst or breech. And because we walk on our hind legs rather than on all fours, there were constraints on how large the birth canal could be. One solution was for human infants to be born early and for their heads and skulls to continue to grow after birth. That meant that they were dependent for a long time after they were born. They could learn a great deal from their mothers in that time, but they needed their mothers if they were to survive.

Every modern parent copes with this human trait day and night, the long dependency of kids in their own knowledge-gathering phases. E. B. White was driving his young stepson to school one day; they were talking about the boy's arithmetic homework when they saw a mother cat and her kittens in the tall grass of a field. The mother was catching a mouse while the kittens looked on. White describes the scene in one of his essays. He says he could not help reflecting how many lessons his son would need to learn before he could go out and catch his first mouse.

Because birth is so hard, and child-rearing goes on for so long, a woman of forty-five or fifty, still fit in many ways, might find it difficult to start all over again as the mother of a new baby. Primitive conditions were so often nasty and brutish. Life was so often cut short. It might have been to her advantage, in the great task of passing on genes, if a woman stopped making babies and began helping her children's children, giving them the fruits and harvests of her lifetime as a knowledge-gatherer. She could do more for her genes as a grandmother.

If you complain to your friends about the aches and pains of getting older, they tell you: Consider the alternative. Aging is hard, but the only other option is worse. The view of the evolution of aging that opened up after Medawar, the view of the disposable soma, is pretty bad, but from some points of view it does suggest a new alternative.

To Medawar and those who followed him it was at least encouraging to think that aging and death are accidents. Aging did not evolve because there was something good about it for the individual or the tribe or the species. Death was not designed. This is the same hopeful message that is written in the Wisdom of Solomon: "For God made not death: Neither hath he pleasure in the destruction of the living." You can look at Medawar's theory as more optimistic than Weismann's. Weismann contemplates the aged, with all their weaknesses and infirmities, and finds it obvious that evolution would have to get rid of them. Evolution has to get rid of them because they are unfit. But why are they unfit? Why do we grow weak as we grow old? After all, as we grow older we grow in experience. In terms of experience, we should be fitter at forty or fifty than we were at twenty. We have wisdom as mushroom-gatherers, say, or mastodon-hunters. We also have immunological "wisdom," which is why parents don't catch as many bugs as their babies. As Medawar puts it, "It must be obvious that, senescence apart, old animals have the advantage of young. For one thing, they are wiser. The Eldest Oyster, we must remember, lived where his juniors perished."

Death is not a punishment for our sins, and aging and death are not designed by Darwin's process either. Aging and dying are not adaptations in the way that our hands, eyes, and brains are adaptations. Aging and dying are misfortunes that visit us because Darwin's process is looking elsewhere, so to speak, busy doing other things. Weismann's argument assumes that we have to decline and fall. That assumption has a lot of evidence behind it; it has the weight of all of human experience behind it; but it is still an assumption. With the problem of mortality, we often do assume what we seek to explain. Aristotle admired nature's wisdom in making our teeth fall out when we get old, because we won't need them when we are dead. The Reverend Thomas Malthus, when he tried to refute the optimistic Marquis de Condorcet on the likelihood of making ourselves immortal, explained death this way: we are mortal "because the invariable experience of all ages has proved the mortality of those materials of which his visible body is made." In the same way, Weismann admired evolution's wisdom in inventing aging as a way of getting rid of the aged. These arguments don't get us very far. Compare the bitter epigram of Jules Renard: "Death is sweet; it delivers us from the fear of death."

If we want to understand why we are mortal we have to learn not to assume that we just have to be mortal. Aging is not an adaptation; aging is just an accident. Death is not made by Darwin's process; it arises because there are places where Darwin's process is powerless to go. Richard Dawkins has called the process of evolution by natural selection the blind watchmaker, because the process creates such intricate machinery without ever looking ahead at what it is making. The forms are made simply by the success of some in each generation and the failure of others—a simple but profound story that we are still in the process of absorbing and digesting a century and a half after the *Origin*. But not only is the watchmaker blind; there is a place the watchmaker cannot reach, a place the watchmaker's fingers cannot touch. That is the desolate place we call old age.

This opens an interesting possibility, a door that we had thought was closed forever. On the old view of aging and death as punishments for our sins, or sacrifices for our children, we could not dream of opening the door without a feeling of enormous guilt and preposterous futility. But on this new view the project of eternal youth and perpetual health becomes as plausible as any other human dream that evolution itself has not granted us but that we might have some hope, with industry and luck, to arrange for ourselves, like flying through the air, or curing the whooping cough, or making life so comfortable that most of us can expect to reach the age of eighty. This view of life suggests the possibility that we might be able to do what the blind watchmaker could not do, and fix and improve, or at least maintain, the clock.

As Medawar observes in "An Unsolved Problem of Biology," we have trouble even imagining that we might get older without declining. "It is," he says, "a curious thing that there is no word in the English language that stands for the mere increase of years; that is, for aging silenced of its overtones of increasing deterioration and decay." Think of test tubes in a laboratory closet, Medawar suggests, or tumblers on a high shelf in the back of a pub. They never get scratched or microscopically chipped along the rims like their siblings out front. Most of them last until they're taken out and put to work, or dropped. Until that fatal day they're practically ageless. In other words, they get older without aging. And so did we when we were children, in a sense—to repeat the point that Francis Bacon makes in the first pages of the *History of Life and Death*. As children we got bigger and stronger every day. Then something happened. And now that we are grown up and falling apart, we are almost as confused about aging as we were when we were children.

Another eminent British medical man of the twentieth century, Robert Platt, a president of the Royal College of Surgeons (later Baron Platt of Grindleford), illustrated this point with an anecdote in his retirement speech in 1963, "Reflections on Aging and Death." Platt said, "The story is told in my family how my brother and I, as small boys, were admitted to an Edwardian tea-party in our house in Hampstead, and my brother in the shrill clear voice of a little boy said, 'Daddy, is Miss So-and-So a young lady?' My father, ever tactful, said: 'Yes, Maurice, of course she's a young lady.' Maurice thought for a few moments and then said, 'She looks as if she's been young for a very long time.'"

In essence, we get old because our ancestors died young. We get old because old age had so little weight in the scales of evolution; because there were never enough Old Ones around to count for much in the scales. But now if we like we can do more to help our bodies in the first half of life so that we will not progress so quickly to the decline of the second—or, in the dream of the greatest optimists, so that they never grow old at all.

And we can begin to do this most efficiently, in the view of those who campaign for the conquest of aging, if we accept the implications of evolutionary theory and come to view aging not as an inevitable and natural process but as a disease. Aging is a disease, like Huntington's. It is just a kind of accident or series of accidents, failures to maintain the body. Aging evolved back when the struggle for existence was more intense than it is for us now, back when we had to race to survive and multiply; when we were much too busy surviving and multiplying to build bodies that would have a chance to last. "Back when" being most of human prehistory and all of living history before that, back to the origin of life almost four billion years ago.

The disposable soma theory helps to explain many confusing things about the problem of mortality. Most important, the theory explains the sheer diversity of the aging process: why so many things go wrong with us as we grow older. That's just what you would expect if evolution cares only about getting you to a certain age; if it doesn't give a damn what happens afterward. In other words, life has a meticulously careful plan for your rise, but no plan at all for your decline and fall. That's why the phenomenon of aging is so hard to explain if you look for a single cause. Aging really has many causes, because none of our myriad working parts was built to last forever. That's why aging is so unlike the orderly development of the embryo. Our development and birth are tightly programmed, but not our deaths. In this sense, the end is not written. It is not written in our stars, and it is not written in our genes.

Once you see the hopeful side of the disposable soma theory, you can begin to put together a grand plan, an escape plan. Maybe—just maybe—we can do for ourselves what evolution has neglected to do. Maybe—just maybe—we can intervene and extend our life spans even more dramatically than they lengthened on the African savannah when we evolved our big brains. Maybe it is now possible to help ourselves to more time—much more time.

Chapter 6

THE GARBAGE CATASTROPHE

"Some scientific discoveries are accepted almost immediately," writes the gerontologist Robin Holliday. The most famous example is the double helix of Watson and Crick. Most biologists agreed within a few years that the two young men really had found the secret of life. Their sprint into the Eagle is now as famous, in scientific circles, as Darwin's voyage of the *Beagle*, or Newton's voyage on strange seas of thought, alone, under an apple tree.

Other great discoveries take decades to be recognized. Alfred Wegener argued in 1910 that continents drift. The idea wasn't generally accepted for more than fifty years. Gregor Mendel published the laws of inheritance in 1866. His discovery was rediscovered after thirty-four years.

Unfortunately, the solution to the problem of aging seems to be falling into this second category, Holliday complains in "Aging Is No Longer an Unsolved Problem in Biology," one of many dozens of triumphant articles, essays, and books that gerontologists have published in recent years. We don't know how to stop it, but we do know why it evolved. In that sense, aging is no longer an unsolved problem. And yet most people and even most scientists haven't heard the answer to one of the deepest and most profound problems that mortals can ask. They haven't heard, or else they haven't understood. "A lot that is written about aging now is biological nonsense," says Holliday, "and that will undoubtedly be true in the future as well."

In the view of the disposable soma theory, aging is simply the slow failure of maintenance. All your life, your body has to keep fixing broken DNA. Clearing away the damage done by free radicals. Repairing proteins. Repelling germs. Detoxifying poisons. Healing wounds. Clotting blood. Mending cracked bones. Adjusting the thermostat to maintain temperature. Adjusting the balance between the destruction and creation of cells to maintain all your working parts, and to prevent a rogue cell from multiplying out of control. Your body does all this internal maintenance work for you as long as you keep up the external maintenance work of eating, excreting, washing, and running a comb through your hair. It takes a lot of work for the body to maintain what it has built, as Holliday notes: about 150 genes just for DNA repair, according to current estimates, and at least a thousand genes for the immune system.

And of course the body has other kinds of work to do besides maintenance. The body invests enormous time and energy into building gonads and attracting a mate to pass on those gametes. And then we put much of our life's energy into feeding and raising the young and helping them grow until they are big enough to go off on their own and maintain themselves.

According to present thinking, it behooves the body to strike the right balance between investing in its own maintenance and in the creation of new young bodies to go out into the world and multiply when it is gone. Because mice rarely live more than a year in the wild but human beings could live for twenty years or more in the wild it made evolutionary sense for the tissues of the two mammals to invest differently. Lymphocytes in the lymph nodes slowly accumulate mutations, for instance, because DNA repair isn't perfect—not in mice or men. In the course of the life spans of both mice and men, these mutations accumulate about tenfold. But they do so in the space of about three years in a mouse, and eighty years in a man. Apparently the mouse doesn't put as much energy into keeping itself up. The mouse lets itself go, as we say, because it is bound to go soon anyway. It makes babies and disappears.

So exactly what would it take to make the human body do even better than eighty years? What would it take to make the human animal immortal? We'd have to be able to regenerate every single one of our working parts, like the hydra, says Holliday. We'd need to be able to rebuild the heart and the blood vessels—without ever shutting it down for repairs. We'd have to repair, regenerate, and rebuild the brain—without losing the memories that make us what we are. We haven't done that because at no stage in human evolution was it ever better and more profitable for a human body to invest its resources that way than to build quickly and pass on its genes.

What we have done instead is to adjust—and fine-tune, generation after generation—the life span of each of our working parts so that they all tend to age at about the same rate. That's why we can look around us and guess the ages of the people around us, according to the disposable soma theory. Our bodies have invested just enough to maintain most of our working parts for the same period, so that they decline and fall at about the same time.

Holliday is one of many gerontologists who believe this theory solves the problem that Medawar first posed more than half a century ago. To Holliday it means that we are never going to be able to live much longer than we do now, because there are so many different kinds of things that go wrong with us that we will never be able to fix them all. So aging is irreversible. Antiaging medicine is a crock. At the end of his review, Holliday quotes Ronald Klatz, who writes in his book *Advances in Anti-Aging Medicine*, "Within the next fifty years or so, assuming an individual can avoid becoming the victim of major trauma or homicide, it is entirely possible that he or she will be able to live virtually forever."

Holliday concludes, with the gloomy air of QED, "This is biological nonsense."

In essence, in the view of the disposable soma, you could say that we come up against a modern form of the legend of the Hydra. Killing the Hydra was one of the twelve labors of Hercules. The monster had nine heads, and she helped guard the way to the Underworld. Hercules couldn't kill her by cutting off her heads with his sword, or his scythe, because each time he lopped off one head, two grew back. He had to lop off every one and cauterize each stump with a torch. Even then he wasn't done, because one of her heads was immortal. He had to bury that hideous head under a rock. And even then, long after he had slain the Hydra, venom from the monster's blood poisoned Hercules, and took the great hero down, wrapped in an intolerable cloak of pain. It was the Hydra that killed him in the end.

Aging is many-headed, like the Hydra. If you are a pessimist, or perhaps a realist, you conclude that you can never kill it. If you are other-minded, you begin to plan your attack.

* * *

The disposable soma theory makes some specific predictions. It predicts, first of all, that aging is caused by the accumulated damage of mistakes in building and repairing the body. The mistakes begin even as the construction begins. We are declining in a sense from the moment we are born. Even from before we are born. From the first moments of the union of the sperm and the egg, we are making mistakes in the hurry to get the building up and get around to the union of more eggs and sperm. As Aristotle said, the smallest error in the laying of foundations can someday bring down a house.

Not long ago I went to visit Janet Sparrow, a medical researcher at Columbia University. She is the Anthony Donn Professor of Ophthalmic Science in the Department of Ophthalmology, with a joint appointment in the Department of Pathology and Cell Biology. In her laboratory, Sparrow is trying to find ways to prevent one of the common vision problems of old age, macular degeneration. It is a simple case of the simplest aging problem, the problem of clearing away debris as we get older.

Macular degeneration is a medical condition that usually begins to develop around the age of fifty. It's a disease of the retina, which is one of those minutely engineered places in the body where you do not want debris to build up. The retina sends the messages to the brain that translate into vision. Our eyesight depends on the health of our retinas, which are extremely thin films of nerve cells at the back of each eyeball.

When a ray of light falls on the rod cells and cone cells in the retina, a certain chemical inside those cells, a chemical derived from vitamin A, has to switch very quickly from one chemical shape to another. The chemical has one shape in the dark and one shape in the light. This switching from the dark form to the light form triggers events that tickle the optical nerve, which sends a message to the brain that a ray of light has arrived. Your whole life, whenever your eyes are open, innumerable molecules of this compound are switching from the dark form (which is known as 11-cis-retinal) to the light form (all-trans-retinal), and back again.

Unfortunately, as it flashes back and forth between its two forms, which is a complicated procedure, one of these molecules sometimes brushes up against one of the molecules around it, and every once in a while the two of them get stuck together. No man is an island, no organ is an island, and no molecule is an island. All of our working parts are working next to hundreds of other working parts. If the wrong molecules happen to brush against each other and stick together, they can begin to clump. In the retina, this molecular accident often ends up as a clump of useless trash, a clunker of a molecule called A2E. The rod and cone cells try to clear away this trash by sweeping it into the lysosomes of cells nearby. But the lysosomes can't break it down. So the A2E sits there inside the lysosomes. After seventy or eighty years of this kind of slow failure, the cells in vital parts of some human retinas are often as much as 20 percent junk: that is, 20 percent A2E by volume. They are almost as bad as cameras that are one-fifth full of dust. This is one of the common problems of old age.

A2E is an ugly and pervasive kind of biological trash called lipofuscin. It's an age pigment. You really don't want lipofuscin in your retinas. When light strikes lipofuscin, it glows, and it goes on glowing for a while even in the dark.

On my visit to Sparrow's lab, I asked her if I could see some lipofuscin. "I'll get a vial and I'll come right back," she said.

The little glass vial she handed me was full of brown muck. She explained that since I was over fifty, my own retinas already contained quite a lot of it. The stuff looked like the kind of crud you get on steel wool when you scour a frying pan.

Meanwhile, of course, all kinds of other material changes are taking place in our eyes as we get older, Sparrow told me. "Have you begun to notice trouble differentiating navy blue and black socks?" she asked.

"Yes, as a matter of fact."

That's a completely different material deterioration, Sparrow said. The lenses of our eyes turn yellow with age. The yellowing is caused by the chemical changes in the lens, and diminishes our ability to see the color blue, because yellow filters out blue light. So as we get older, we see blue less brightly. Often, people who undergo cataract surgery to have a cloudy, yellowed lens removed and replaced by a clear new artificial lens can suddenly see all the blue light they experienced sixty or seventy years before. "Patients say, 'Oh, blues are so bright. The sky is so blue! I haven't seen that blue since I was a child!'"

The yellowing of the lens has nothing to do with lipofuscin. Neither does still another kind of junk, called drusen, which eye doctors can see glittering in the back of an aging eyeball through an ophthalmoscope. Drusen looks through the scope like tiny, shiny crystalline dots, whitish and yellowish. The term comes from the German word for a geode: drusen resembles the cup of semiprecious crystals you find when you split open a geode. Eye specialists have known about drusen for more than a century without being able to figure out where the crystals come from or whether they're early-warning signs of macular degeneration. Drusen crystals also start showing up around the age of fifty.

That's the way it is throughout the body. You get rare, semiprecious, specialized kinds of junk like drusen crystals in the eyeball, or crystals of calcium oxalate in the kidneys, which are called kidney stones. Other kinds of junk are found throughout aging bodies and can turn up almost anywhere, like lost sheets of newspaper, cigarette butts in gutters, plastic bags in trees, crumpled tissues in wastebaskets. Lipofuscin piles up in cells in many parts of the aging body, but seems to accumulate most in cells that do not divide. Skin cells and the cells that line our guts are always dividing and being sloughed off. Not much trash builds up in them before they die and are replaced. But heart cells and nerve cells have to last us our whole lives. About 10 percent of the mass of the heart of a centenarian is lipofuscin.

After making his study of aging mitochondria, Aubrey de Grey was fascinated by this problem of the accumulation of junk, rust, and scrap in the body. If that's all aging really is, the slow accumulation of damage, then it's reasonable to argue that there are three ways to fix it. You can try to repair our metabolism so that it does not generate so much trash; you can try to clean up the trash itself; or you can try to deal with the harm the trash does to the body. That's when the problem passes into the domain of surgeons and geriatricians and home health aides. They help elderly patients with their weakening muscles, weakening eyes, cloudy lenses, stiffening joints, wrinkling skin, thinning hair, rusting memories, and on and on, the whole lugubrious list of symptoms that we all know from the inside out, and have always assumed that every generation that follows ours will have to endure as we do.

Aubrey decided that the easiest place to attack the problem is in the middle, in the cleaning up of the trash. The beginnings are too complicated. Metabolism consists of too many interconnected networks for anyone to safely intervene. It's almost beyond imagination how complicated and delicate the action is in the retina when light strikes. And when you eat a piece of food, and some of those nutrients reach a cell in your skin, all of the networks of genes in your cell that have to work together to turn that bit of nutrient into a bit of you are unimaginably complicated, too. In the immortal verse of Walter de la Mare,

> *It's a very odd thing—*
> *As odd as odd can be—*
> *That whatever Miss T. eats*
> *Turns into Miss T.*

Metabolism is a terribly complicated thing as well as odd, and to try to intervene in all of those invisible molecular pathways would make trouble for Miss T.

The pathologies of old age are also complicated, and—as anyone knows who is in the middle of them, or has watched a loved one endure them—they are interconnected. If macular degeneration is allowed to progress untreated it causes incurable blindness. In the Western world, it is now the most common cause of incurable blindness. Before you reach fifty, it's rare, but by the time you pass eighty, the incidence is one in ten. Once Miss T.'s retinas are damaged, she is more likely to fall; once her bones have grown frail because of osteoporosis, she is more likely to break her pelvis when she falls. She is often dizzy anyway, and she has lost some of the redundant systems of nerves that used to help her keep her balance. Meanwhile, because of the osteoporosis, the vertebrae in her lower back are painfully compressed. When she breaks her pelvis, her surgeon finds it hard to operate on the vertebrae; and on and on. Meanwhile more damage keeps piling up.

But the dirt itself, the little piles of dust and lipofuscin and miscellaneous debris that accumulate in the corners and crannies of the cells, and cause the damage—that is comparatively simple to deal with, in Aubrey's view. Cleaning it up may be hard, of course. But it should be easier to clean up the dirt than to overhaul the entire industrial landscape of the body, which produces the pollution; or to repair the body as it falls apart at last and Miss T. breaks down and dies.

Aubrey began seeking out researchers who study the pollution and ways to clean it up. On trips to Boston, Aubrey visited Ana Maria Cuervo, who was then in training at Tufts University and now runs a gerontology laboratory at the Albert Einstein College of Medicine, in the Bronx.

Cuervo studies the action of lysosomes throughout the body. The lysosome is the organ of self-sacrifice, within the cell. With its lysosomes the body does unto itself what it does unto others. Chomp,

chomp, chomp. Producing nutrients to digest and recycle, along with a kind of microscopic excrement, indigestible trash.

Cuervo has done as much as anyone to show that the body gets weaker and weaker at taking out garbage as we get older. She has been working on garbage and lysosomes ever since she started out working toward a Ph.D. in the early 1990s. The lysosomes have fascinated her all her working life. ("Such a little fellow, and so much to offer.") Cuervo wants to understand the ways the body carries out its continuous acts of self-immolation, in which not only old mitochondria are carted off to be scrapped but virtually every bit of the cell is perpetually dismantled and recycled for spare parts and reassembled.

Ana Maria Cuervo and Aubrey de Grey are friends—an odd pair of friends. He lives on beer and she lives on Diet Coke. She keeps a shelf of Diet Coke cans from all over the world above her desk and she's always pouring another plastic cup of Diet Coke for herself and her guests. She is one of the few people I've ever met who talks faster than Aubrey. She agrees with him that the key to the problem of aging may well lie in a kind of sophisticated detoxification of our cells. She's an experimentalist who hopes we can make cells live a long, long time by giving them extra genes for taking out the garbage. She writes papers with titles like "Keeping That Old Broom Working" and "The Ultimate Cleansing Diet." In their campaign to figure out how to detoxify the body, to take out the garbage, she and Aubrey are comrades in arms. But when he talks about immortality, she just laughs. Immortality has absolutely no appeal for her. English is not her first language (she comes from Barcelona) and for some reason she always calls him Audrey.

"Audrey," she says, "if I have to be here five thousand years, take me now!"

In her laboratory, Cuervo is trying to understand the molecular action in the cell's chop shop. The lysosome is constantly devouring the rest of the cell and cutting it up into recyclable pieces and exporting those pieces to be reassembled in its daily acts of renewal.

Until recently, few people were interested in aging lysosomes. Aubrey was ahead of the curve in focusing on them and befriending Cuervo. She and her handful of fellow researchers worked in obscurity. Science has fashions, and elderly lysosomes were unfashionable. The genes that control the pathways by which the cell sweeps and carts bits of itself into the lysosome are known as "housekeeping" genes, and almost nobody was excited about housekeeping genes. Two biologists at the NIH, Shiwei Song and Toren Finkel, began a paper on housekeeping genes with a complaint about their low status. "In most schools, students tend to stratify into groups like the cool kids and the nerds," Song and Finkel wrote. In the genome, too, some genes seem to get all the attention, and "life is a lot less glamorous for everybody else." At the bottom of the "uncool" list were housekeeping genes.

If lysosomes were called the nest of the Phoenix, instead of the trash can, they might have more glamour. Of course, the whole field of gerontology suffers from the same image problem. It is seen as the science of aging, and most scientists in their prime find the thought of aging as unattractive as the thought of housekeeping.

But housekeeping is an inadequate description of the magical act of self-renewal on which life depends day by day; and aging is an inadequate description of the mysterious way in which this act of renewal slips and declines, very gradually, day by day. And this is not a corner of our existence; this is what we are. When we talk about eating, and taking out the trash, we are talking about fantastically complicated acts of creation and destruction. We are talking about the ways our bodies, some of the most complicated things in the known universe, destroy and rebuild themselves daily and hourly.

Humble housekeeping genes help cells divide and develop. They help cells fight off invading bacteria and viruses; they help the immune system. When something goes wrong with all this housekeeping we can develop cancer; or neurodegenerative diseases like Alzheimer's and Parkinson's. One of the secrets of success, we say, is just showing up; and one of the secrets of staying alive is just housekeeping.

Cuervo set out to learn if the decline in housekeeping is a major cause, or even the major cause, of aging. Clearly the decline makes the cell less efficient, which means that the cell produces more trash and cleans up less. If this failure of housekeeping leads to our mortality, then in effect we die from a pileup of junk.

The level of fine detail with which Cuervo and others can watch all this mortal housekeeping is amazing. Maria Rudzinska could only stare at a cell with a microscope as it grew old and filled with strange dark particles and died. She could no more see molecules than a tourist at the top of a skyscraper can see pebbles. Now Cuervo has the benefit of the half a century of tools that have been invented since Watson and Crick opened up the exploration of molecular reality. The tools with which Cuervo's generation watches the action include not only light microscopes and electron microscopes but also special stains that make the working parts of molecular machines light up and glow in living cells as if they were followed by Broadway spotlights; along with all kinds of tricks that she and others have developed, including centrifugation of cell corpses through dense cushions, and the use of fluorescent dyes that stain certain streams of the autophagic traffic red and blue. Tricks of genetic dissection and X-ray crystallography allow them to open up molecular machines and count the teeth on each gear.

By the late 1990s, Cuervo and other specialists had identified five different pathways by which the cell keeps house. Sometimes a lysosome digests a big chunk of the cell around it. This is known as macroautophagy—literally, consuming yourself in big bites. Sometimes the lysosome chews a smaller bite—microautophagy. And sometimes the lysosome takes in a single molecule, which is the smallest nibble possible and requires remarkable precision, like picking up a single grain of rice with chopsticks. The lysosome uses a sort of claw hand to seize and grasp individual molecules for engulfment and dismantlement. The claw hand grabs the molecule of junk and holds it while other lysosomal machinery unfolds and unspools it into a long loose ribbon. Then the ribbon is drawn into the lysosome like a sheet through a porthole, yanked or pushed and wadded through by still other molecular machines of a class known as "chaperones." There are chaperones on both sides of the porthole. Some chaperones wad the sheet through the hole from the outside, and some of them yank it in from the inside.

From the beginning, Cuervo was most interested in the ways that lysosomes might be involved in aging. By the year 2000, Cuervo and others had shown that most of the pathways by which the living cell carts bits of itself to the lysosomes for demolition and recycling do decline with age. The cells of young laboratory rats work twice as hard as old rats' at carting bits of their cells to their lysosomes. The lysosomes in old cells grow swollen and frail. They fill with aging pigments, along with other indigestible junk, including the stuff that free radicals make as they carom into molecules inside the cell, leaving them tangled and cross-linked in ways the lysosomes can't cleave and dismantle. One of the simplest kinds of detritus that accumulates in our bodies is the kind that makes our skin wrinkle. In the United States alone the market for what are now called "cosmeceuticals" is more than $8 billion a year—for ointments that try to do what the first ointment claimed to do on the banks of the Nile in 1500 B.C. And the cause of our wrinkles is such a very simple thing. What makes our skin supple and smooth is a protein called collagen, as anyone knows who has read the ads and the labels of the antiwrinkle ointments. Each collagen protein is a molecule shaped something like a long rope. The ropes are very strong and they are arranged in the skin in great woven nets, something like the nets of rope baskets. Unlike rope baskets, however, they are alive. As part of our living bodies, part of the bright burning life of the Phoenix, they are continually made and destroyed. Unfortunately, as time goes on, they are made less well, less accurately. They tangle with each other at the edges. They collect what are known as cross-links, which are tiny ties that join one rope to its neighboring rope and stiffen the whole net. This is what makes our skin stiffen and wrinkle—and inside our bodies, too, daily, nightly, in each of our internal organs, in our arteries and veins, in the kidneys, the liver, the eyes, the brain, the same unfortunate cross-linking goes on, with results that can be much more serious than wrinkles. These cross-links are known in the jargon as advanced glycation endproducts (AGEs).

When we're young we have a spring in our step, as we say. Actually, we have a million springs that put that bounce in our steps. Picture what would happen to a spring or a Slinky over time if you stapled more and more of its coils together, at random. Eventually the body is not very springy, or slinky, anymore.

It may also be that the cell, as it gets older, grows less nimble and deft at the crucial folding operations that produce the elegant origami of its molecules in the first place, so that there is more and more crumpled, badly folded trash for the lysosome to handle and dispose of. Chaperones inside the cell are actually able to decide if a given bit of origami is close enough to right in its folds to be worth fiddling with, or if the whole thing is such a botch that it would be better off chucked. And if the cell can't make enough well-folded origami, and if the lysosome can't split the stuff up and spit it back out to be recycled, then the cell has less raw material with which to try again with fresh origami. So the cell begins to weaken at both self-creation and self-destruction. The one can't suffer without hurting the other. You can't be creative without tools, fuel, and quality control. To make good things, you have to throw bad things out. As Isaac Bashevis Singer once said, a writer's best friend is the wastebasket.

Aubrey thought about housekeeping genes, and he thought about all the commonplace trash that escapes the broom and gathers in the corners, like molecules of lipofuscin, which litter billions of aging cells like little balls of dust. Although lipofuscin is a confusing substance to biologists, it is clearly related to metabolism, since it accumulates inside our lysosomes as we get older. "That says it's not doing any good," Aubrey says. "Even lysosomes can't break it down. In spite of having about sixty different enzymes to break things down." Lipofuscin is like the dust in the corner that the blunt broom can't get at. It is like a spoon or fork in the garbage disposal, or the wad of glop in the S curve of a drain.

Eventually Aubrey hit on a way to deal with this particular garbage problem. The idea came to him in an epiphany. It was the end of a long day's journey with a duffel bag. He was attending the 1999 Annual Meeting of the Society for Free Radical Research, in Dresden. Aubrey had already been brooding about the kinds of specialized shears and scissors that might help to cut the cross-links in AGEs, the kind of junk that gives us wrinkles. Suddenly in Dresden it struck him, he says, that all of this junk is cross-linked. All of it is the by-product of metabolism. All of it is crazed and crumpled molecular origami. The body's problem throughout—in the skin, the heart, the nerves—is that it has never evolved the proper tools for uncrumpling the most tightly wadded sheets, snipping the most tangled tangles, the toughest chains in the cross-links. And the reason the body has not evolved the tools is that it has not needed to do so. All this junk accumulates gradually, on the whole. It is not a problem for the body until midlife or beyond.

And it dawned on Aubrey that he knew the man to fix the problem. He knew a good man in the genetics department in Cambridge, John Archer. Archer searches for soil microbes that can devour toxins that we are unable to seek out and destroy ourselves. He is a specialist in the field known as bioremediation, or environmental biotechnology. It is becoming possible in some cases to decontaminate soil that has been poisoned by dioxins: bioremediation experts have genetically engineered microbes that eat dioxin. Likewise, they have ways to clean water of PCPs, using still other poison-loving microbes. All of these pollutants, bizarrely, can be broken down by microbes.

Rubber! Go to the side of the highway. Very tiny bits of rubber are continually flayed from spinning tires of cars and accumulating on the side of the highway. Specialized microbes have evolved there in the speeding shadows of our cars and trucks and they feast on the rubber dust. You can find bacteria there that will eat it. If you are looking for enzymes to dispose of rubber, that's a good place, among the microbes of the roadsides. The microbes have already found the answer—so you can collect them from the grunge at the side of the road, and raise them in petri dishes, and study all of the tricks for the disposal of rubber that evolution has discovered.

If roadsides are the places to look for the secrets of the disposal of rubber, then where are the places to look when you are concerned with the disposable soma—when your problem is the decades of accumulated trash in each and every human, mortal, disposable body? Where have we mortals disposed of this tragic debris for generation upon generation upon generation?

Graveyards.

Aubrey's hometown is rich in old graveyards, including Coldham's Common, where the people of Cambridge buried many generations of their dead, including corpses from their leper colony in the twelfth century, along with victims of the Black Death; and in the seventeenth century, the Great Plague. Midsummer Common is another hoary Cambridge graveyard. Aubrey thought of the planet's most notorious mass graves, from Rwanda to Cambodia to Dresden itself.

In the meeting that day in Dresden, one of the presenters was Ulf T. Brunk, chair of pathology at Sweden's University of Linköping. At the meeting in Dresden, Brunk showed slide after slide of elderly cells clogged with lipofuscin. The stuff glowed a dull red on his slides. Lipofuscin is Brunk's specialty.

Aubrey was sipping coffee after Brunk's talk when the thought came to him, and he hurried across the conference room.

"Listen, Ulf, I've just had the most fabulous idea . . ."

Ulf's reaction disappointed Aubrey. Ulf seemed rather cool about the idea of prospecting for a cure for aging in graveyards. Aubrey put it down to Nordic caution.

After he got home to Cambridge, Aubrey shared the idea with Archer, who saw Aubrey's point immediately. Archer summed it up in a single line: "Why don't graveyards glow in the dark?" So many centuries of lipofuscin-laden remains have been buried there. They are the repositories of all of the tangled, mangled, ruined molecules that we the living (while we were still aboveground) had never quite managed to dispose of ourselves. And yet the soil of our graveyards does not fluoresce. After eighty, our retinas are fluorescent too, because of their loads of lipofuscin. Graveyards should be full of the stuff, and yet they do not glow. Obviously, microbes in the soil must have found ways and means to work through the coffin-lids and the winding sheets and cerements and devour the very last of the debris. After all, our bones are picked clean by the scavengers in the soils of graveyards. They whittle us down to skeletons as we rot. So Aubrey proposed that we dig in the old graveyards and look for the secret in the bacteria that have evolved there. Steal the tools from the Lord of the Underworld, from the Devil's workshop.

Aubrey was not the first gerontologist to follow his thoughts a little farther into the grave than most of us like to go. Medawar had gone there before him. In his essay "Old Age and Natural Death," Medawar talks about the difficulty of defining the moment of death. He points out that because we are made up of trillions of tiny living cells, some of them are bound to survive us for a long time after the doctors pronounce us dead, "and those whose most pressing fear it is that they will be lowered living into their graves can have their doubts resolved: they will be."

This was the way Aubrey did science. He worked away in the Department of Genetics, like a newfangled scribe, on a great compendium called FlyBase, entering lines of computer code that define the genes and mutations of the fruit fly. Now and then he went dashing up and down the old spiral stairs, elvishly, to visit Archer, or to drop in on his wife, Adelaide, at her tiny lab under a stairway leading to the roof, and try out a new idea.

"Well, I mean, it's deucedly simple, really," John Archer said, when Aubrey and I dropped in on him at his lab in the summer of 2004. Computers hummed on the desktop. Archer was running genomes. That took a lot of computer power. So the little room was warm, in spite of the labors of an expensive air conditioner of the same make and model as the ones that cool the sealed capsules of the London Eye, also known as the Millennium Wheel. Archer wore an Izod shirt, khakis, an ID around his neck. Everything about him said solid, a man with his feet planted on terra firma. Whereas Aubrey had, as always, that millennial hippie look. He wore a red "Drosophila" T-shirt over a flowered Hawaiian shirt, with a red elastic for his ponytail.

Archer is an expert on the metabolism of explosives. He explained to me that soil around military camps may be contaminated with nuggets of TNT as big as an inch in diameter. When rainwater pools on ground that is loaded with TNT, it turns a telltale pink. To cure the soil of these explosives and poisons, some specialists work on sowing these fields with plants, including the wild tomato and the downy thorn apple, known in the Wild West as jimsonweed. Jimsonweed will take the pink out of a jar of pink water overnight. Some army bases now try to keep their poisons from seeping downward into the groundwater by planting trees that suck them upward.

Archer prefers to work with a kind of bacteria called Rhodococcus, which is a master at the digestion of TNT. Archer feeds it pink water and watches what happens. ("Get a little drop of that on you and you're going to have an interesting time at Heathrow," Archer said.) Strains of Rhodococcus are robust and diverse and can eat up a wonderfully wide variety of explosives, poisons, and potions, including quinolone, some particularly stubborn thiocarbamate herbicides, and a chemical called 2-mercaptobenzothiazole, which is used in the manufacture of vulcanized rubber. Strains of Rhodococcus are the only bacteria known to eat benzothiophene and dibenzothiophene. They thrive in ethanol. Archer is proud of Rhodococcus, of its hardihood and its appetites. He keeps collections of polluted waters ("It's wicked, wicked stuff!") and he keeps potent antidote strains of Rhodococcus. He likes to talk like a plumber—he says that's what he would probably have been a few centuries ago, an engineer tending the drains of some lord's manor house. "When these first started growing, they et the petri dish," Archer said, in mock-peasant style, pointing at one of his favorite strains. "Et it right up. They melted the petri dish and drank the solvent." He boasted about the Rhodococcal metabolism. Our bodies can't do a tenth of what his Rhodococcus can do when it comes to digesting explosives. "Humans are pathetic," Archer said. We may be more complicated than bacteria, he said, "but metabolically we're crap."

Bacteria are so gifted metabolically because they are far more diverse than we are, Archer explained. Four-point-eight billion years ago this planet was nothing but rock. Four and a half billion years ago, the planet came to life. Most life-forms are still devoted to being unicellular. Bacteria are phenomenally diverse. They made the world, and they will inherit it—they'll still be here when we're gone. "I mean, my God, you know we're really Johnny-come-lately," Archer said. "We just got here." Bacteria have been fine-tuning their DNA for four and a half billion years.

"Soil is more active than you'd ever realize," said Archer in a hushed voice. "Tremendous energy."

Archer was very impressed by Aubrey's energy, too, but he hadn't done much with his graveyard idea. He had gone as far as to send a student to Midsummer Common with a trowel. There the bodies of plague victims of Cambridge have lain moldering for centuries. And he had done a preliminary experiment. Beyond that the soil samples from the graveyard were still sitting around in his refrigerator. Aubrey's idea was only one of many projects that Archer could take up, and he did not seem in a hurry to do much with this one, when he had so many other experiments of his own to try.

This is one of the hazards of being a theoretical biologist like Aubrey. You have to interest people with laboratories to take up your ideas, and they tend to have ideas and projects of their own. Hands-on biologists often remind Aubrey how hard it is to make things work in the laboratory. They're inching up the mountain while Aubrey dreams of landing up there in one stride with his seven-league boots.

They tell Aubrey, "It's a grim life doing experiments."

"C'est la vie," says Aubrey.

From Adelaide, Aubrey might have learned some humility about the difficulty of doing experiments. Years had passed since her brilliant start as a geneticist, her discovery of a particle within life's machinery; which had looked to her at first in her slides like a speck of dirt. That particle of molecular machinery helps to shuffle the genes in flies, mice, oaks, and people. She named it the recombination nodule. She was still famous in some quarters. But now, after that early success, she spent months or sometimes as much as a year as a technician in the genetics building, trying to unravel what had gone wrong with someone else's experiments—because experiments don't always work out. She labored under the eaves on the top floor of the Department of Genetics. She had a little nook beneath the stairs, with books stacked under the stairs, along with piles on piles of boxes, equipment, papers, books, tissues, yellow pads, pens, pencils, lightbulbs, a crumpled scarf, and an old brass magnifying glass with a wooden handle. Her desk was a few steps away from her fruit fly emporium, with a microscope, and an etherizer, everything she needs for breeding and sorting flies, and fixing the experiments of students whose work refused to go—projects that were failing to progress in anyone else's hands.

Computer programs can also provide lessons in humility. Aubrey's friend Aaron Turner was still struggling toward their dream of the Cure-All, the computer program that would cure all other computer programs. When their branch of Sinclair Research was sold, Aubrey and Aaron had created a two-man company they called Man-Made-Minions and begun racing to develop the Cure-All. In the beginning of their collaboration, they often met at the Eagle or the Live & Let Live for pool, with beer (for Aubrey) and Coke (for Aaron), and talked of all things AI. Now Aubrey was saving the world from death (whether or not the world wanted saving). Poor Aaron was living out of his car, and quoting sardonic sayings of the computer trade, such as Hofstadter's Law: "It always takes longer than you expect, even when you take into account Hofstadter's Law." Aaron slogged away at the prototype of the Cure-All, with moral support and occasional cash from Aubrey, who still served as codirector of Man-Made-Minions. Aaron has since written a memoir of the Cure-All adventure. The final section of his memoir is titled, "Tomorrow and Tomorrow and Tomorrow Creeps in This Petty Pace from Day to Day."

Nevertheless, both Adelaide and Aaron indulged Aubrey's hopes of the Cure-All of Cure-Alls. They admired him. Someday it might be Aubrey's turn to have his words on display in the Eagle, like James Watson and Francis Crick, when he had found the secret of eternal life. Someday it might be Aubrey's turn, like the young pilots of the Battle of Britain, to place a chair on a table, stand on the chair, and sign or singe his extraordinary name across the ceiling. Justify your existence!

John Archer told me that he found Aubrey's idea provocative enough to be worth taking up in an experiment in his laboratory—not as a major experiment, but a Friday experiment, as he put it. It seemed to me that Archer was almost as uncertain about Aubrey's ideas as I was, although he spoke about him with great warmth and enthusiasm. "Now, Aubrey has phenomenal energy," Archer said, "and the trouble with science is—" He plunged into a colorful tirade about the faults and conservatisms of scientists, who can't see past their own noses, and who eat their young. Here you have a bright young man of spirit, a newcomer whose views are just a wee bit novel, and what do his elders choose to do but gather round and devour the poor lad alive. "In reality we need more Aubreys, God help us. Chaps like him who can see over the hedges. What we've got here is a Renaissance man. Missed his century by a few hundred years."

Since then, by dint of persistence and propaganda, Aubrey has managed to get some high-level scientists to take up his graveyard idea, the microbial remediation of the aging body. There is a bioremediation program exploring Aubrey's idea at Rice University; and another at Arizona State University, led by Bruce Rittmann, one of the world's leaders in bioengineering and bioremediation. Rittmann is an authority on breeding bacteria to clean up Superfund sites. On the side he's now trying to clean up aging cells.

The field as a whole—the study of good housekeeping as the key to good aging—has grown a great deal in the years since Aubrey began championing it. When old cells begin to fumble with their molecules, whether in the course of construction or destruction, creation or demolition, they can accidentally make junk that is bad for them. A certain protein that most cells manufacture daily for use in their membranes, a protein as commonplace and crucial in the membranes as plywood sheeting in the walls of construction sites (although nobody knows what it is for) can get accidentally misfolded to form the junk called beta-amyloid that accumulates between the brain cells of people with Alzheimer's disease. Beta-amyloid is badly handled by those garbage-disposal units the lysosomes. It builds up in the brain in deposits like drifts and junk heaps. Trouble with lysosomes may be a cause of the buildup. Failing lysosomes may also help hasten late-onset diseases like diabetes, thyroid problems, and the weakening of the immune system. And of course the aging pigment lipofuscin also builds up in old retinas and causes macular degeneration.

Confusing things can happen in lysosomes, as in the fog of war. Cells can use them to devour invading parasites like the bacteria streptococcus. On the other hand, a cell that has been hurt or poisoned may devour a good part of its own substance and die. Nobody knows whether the cell is trying to kill itself or heal itself. Apparently the very same mechanism can promote survival and promote death. Perhaps when the cell devours itself too fast to rebuild, it dies.

It may be that our ability to maintain a healthy balance between the creation and destruction of molecules is what makes the difference, at the very finest scale of mortality, where we are examining life and death virtually one molecule at a time. Specialists in the lysosome, along with another cellular disposal unit, a stout, barrel-shaped structure called the proteasome, like to argue that housekeeping may turn out to be at the heart of it all. Many gerontologists think the Good Housekeeping people are a bit too enthusiastic; they think the self-cleaning, self-devouring work of autophagy is only a part of the problem of mortality. But it is certainly true that autophagy plays a role during the body's time of growth, from embryo through youth to maturity. It is one of the ways the body sculptures itself to produce its final form, carving away webs between the fingers of the young embryo to produce the hand, or whittling away excess neurons from the brain of the infant to produce and refine each working mind. The machinery of autophagy is also important when aliens invade; when bacteria and viruses intrude on the body, some of the defensive work of demolition is done by autophagy. And autophagy is crucial at every moment of our lives in the nest of the Phoenix, where we are continually consumed and reborn. But our bodies are not designed to do it perfectly forever because our whole bodies are, in the last analysis, disposable.

It is easy to see how trouble in lysosomes might spiral out of control as we get older. For instance, free radical damage may interfere with the lysosome's ability to digest big bites of the cell—macroautophagy. Then, because the lysosome can't handle those big bites of the cell, more free radical damage builds up around it. When a cell is young, these bites really are gigantic. A healthy young lysosome can swallow a mitochondrion—it can take in a whole factory in one gulp. Because an aged lysosome can't do that, more mitochondria that need scrapping may sit around unscrapped. And because the cell can't swallow large chunks of itself at one go, billions of smaller molecular machines inside the cell have a longer "dwell time." They sit around longer, increasing the chances that they will get bunged up and malfunction and make other stuff badly, which will then sit around, too, mucking things up.

In Sweden, Ulf Brunk and Alexei Terman, a colleague of Brunk's at the University of Linköping, have coined a name for the hypothesis that it is junk that gets us, trash that brings us down. In their hypothesis, the more garbage there is in the cell, the less efficient its metabolism. As all good housekeepers know, each bit of junk you leave lying around makes it more likely that there will be more junk lying on top of it or to the side of it, crap piling on crap. At last you get an explosion of junk, according to Brunk and Terman.

They call this the Garbage Catastrophe.

Chapter 7

THE SEVEN DEADLY THINGS

In the summer of 2000, Aubrey de Grey was invited to give a talk at a meeting in Los Angeles focusing on what to do about aging and how to bring all these lines of work together into a single research program. At the Marriott Hotel in Manhattan Beach, speaker after speaker presented hourlong talks reviewing one aspect of the aging problem or another. Each speaker analyzed one line of work without much reference to the others. As he listened, Aubrey felt demoralized. This really is a public-relations problem for the science of aging, he thought: how scattered and incoherent it all is. Aging is so chaotic, so Hydra-like. No wonder the world despairs of a cure. So many monstrous jaws agape, so many terrors gnashing their teeth at us all at once. And each gerontologist fights one set of jaws and ignores the rest.

Aubrey went to sleep in his hotel room feeling exasperated. Because of his frustration and his jetlag, he woke up after only a few hours. It was four o'clock in the morning—or noon, back home in England. Aubrey remembers sitting up in bed, tugging at his beard, and seething about the day. Suddenly a thought came to him: Why not just clean up *all* of the junk? If aging has no program, no plan, if we just fall apart in slow random motion, then there's always going to be chaos in each body's decline and fall. The evolutionary theory of aging predicts chaos. And chaos is just what you see at the cellular and molecular level, and what you will always see. But what these troubles all have in common is that they fill the aging body with junk. Maybe we can just clean up all the scree and rubble that gathers in our aging bodies. That is what came to him in the hotel room in California at four in the morning.

Aubrey told me all this a few years later. He was passing through the United States on a lecture tour in 2002, and I invited him to stop by for a day or two to explain what he was about. It was our first meeting. I picked him up at the airport in Philadelphia and as he huddled in the death seat of my car, enduring the curves of our country roads (we were winding our way toward my house in Bucks County, Pennsylvania, where I lived at the time), the poor man's face looked pale to the roots of his hair and beard, corpse-gray in the cheeks. His head tilted toward the window as if toward martyrdom. He'd never learned to drive, he told me, when I asked him if he was all right. He couldn't help feeling frightened in cars. That's how everyone will feel, he said, when they realize that they may have hundreds or even thousands of years ahead of them. Once that truth sinks in, it will be hard to find anyone on Earth who is willing to ride in an automobile, much less a police car or a fire truck. Too much life ahead. Too much time to lose.

We stopped at a country tavern, the Pineville, and Aubrey relaxed with a beer or two. Once we were settled in my study and Aubrey was relaxing with another beer, I asked him about metabolism. To have any chance at engineering longer lives, didn't we have to understand the intricacies of metabolism—all the ins and outs of our building up and our tearing down, all the invisible labors of the Phoenix, for which we burn away each day at 98.6 degrees?

Aubrey explained the thesis he'd been developing for the past few years. "We don't have to understand metabolism, because we don't have to clean up metabolism," he told me, triumphantly. "All we have to clean up is the detritus that metabolism lays down. And the critical point is, the detritus is not complicated at all. There are only seven types of detritus, more or less. This is the key insight that underpins everything I say. It hasn't been thought about by gerontologists. They're scientists, not engineers."

I had spent enough time in pathology labs, staring through microscopes at damaged tissues that reeked of formaldehyde, to know that our mortal detritus is incredibly complicated. But I'd promised to listen with an open mind.

Aubrey asked me to recall the theory of the disposable soma. After the age of reproduction, the whole body is disposable; so the garbage piles up. It is all very simply a problem of garbage disposal. Our bodies were not designed to last as long as we would like them to last. So why not keep them in good repair, as you would keep a treasured antique car in repair, scraping away the rust, replacing broken parts, and so on? We maintain our houses, too. If we want them to stay leakproof we have to caulk the sills, replace the roof every ten years. We have to repaint. We have to replant, reseed the gardens, clean the gutters, repoint the bricks with fresh mortar. It's a lot of work, of course, but there's nothing very mysterious or surprising about it. If we do all that and keep after it, the house will last a long time. So why not do the same thing for our bodies?

The beauty of this view of aging, Aubrey said, the beauty of the disposable soma, and the garbage catastrophe, is that curing aging requires no great knowledge of design. You don't have to be able to design a car in order to maintain a car, or to build a house to maintain it, either. "Most people want to clean up metabolism—and metabolism is so crazily complicated. It won't happen," he says. "It'll be decades before we understand the cell like a nuclear power station. So my radical idea is, don't try to prevent damage. Let it take place." And just keep cleaning up after ourselves—keep clearing away the nuclear debris. Although metabolism is complicated, the nasty byproducts are not. It's like the difference between a car and the rust and gunk in the engine. The mess may be hard to clean, but cleaning it out is a lot simpler than designing and building a new engine. Rust and gunk have no working parts. They just sit there and get in the way, he said.

Aubrey laid out this argument that day with prodigious intensity, leaning or half-lunging out of my office chair, bottle in hand, the beard pooling in his lap. Behind him stood my wall of books about biology. By this time I had quite a collection, after two decades of following the science of life, a long tall wall that ran the length of the room. I enjoyed Aubrey's enthusiasm and I thought he was a great character, but I wasn't hearing the kind of trumpet blast that would bring that wall of books tumbling down. Probably I was giving Aubrey the hairy eyeball now and then as I listened to him; but he was used to that, and that's why he talked like an overeager salesman. With his larger-than-life beard and intensity he reminded me of a prisoner in a cell in an old *New Yorker* cartoon. Two men hang chained to opposite walls of a medieval dungeon. Nothing in the cell but tall stone walls, one tiny thick-barred window. Each wretch is chained about ten feet up on the wall, spread-eagled by shackles on both wrists and both ankles, each man with a beard of years hanging down past his waist. And the first prisoner is saying to the other, "Now, here's my plan . . ."

With that kind of confidence, which struck me as slightly ludicrous, in the face of our mortal situation, Aubrey told me the story he has told a thousand times since: how sitting in his hotel room in the small hours that night, he made a list of the kinds of junk that builds up and what we might do to clear it away. When he was finished with his list he was greatly encouraged to see that it was short. Once you think of our body's decline in terms of clearing away just a few types of damage, you've demystified the problem of the ages. You've reduced the greatest prisoner's dilemma of human experience to nothing more than a list of puzzles. With our big brains, we are wonderful at solving puzzles and brain-teasers. There's nothing that a little can-do spirit can't do if we all pull together. I listened to Aubrey and I thought: There it is again, the voice of immortalists down through the ages. It was the same ageless voice I first heard in the Reading Room of the great public library on Fifth Avenue, up the flight of stone stairs between the lions Patience and Fortitude, when I met Ko Hung, Roger Bacon, and Paracelsus (born Theophrastus Phillippus Aureolus Bombastus von Hohenheim).

The list had changed a bit over time. He'd started out with nine deadly things, in his first paper on the subject. Since then he'd been adding and subtracting, lumping and splitting, until he had seven.

All seven are well known, even notorious. In fact, Aubrey's seven deadly things are to gerontologists what the seven deadly sins are to doctors of the soul. Here is one deadly thing: many of our molecules grow tangled and stiff with age; they get stuck together in more and more places, as if demons were dashing through our bodies daily with molecular staple guns. Biochemists call those staples cross-links. Here is a second deadly thing: the cells' mitochondria fail with age. A third: junk collects inside our cells. A fourth: more junk collects in the spaces between our cells. A fifth: some of our cells get old and hang around in the body without doing their jobs—making a nuisance of themselves. A sixth: some cells die and poison the cells around them. And here is a seventh deadly thing. The very worst citizens among our cells accumulate dangerous mutations in the genes of the nucleus. Those cells' descendants build tumors.

Aubrey explained that if we approach these seven deadly things methodically, the conquest of aging might turn out to be a very simple matter, at bottom. Start with the most basic kind of damage, the cross-links that stiffen our collagen and make our skin wrinkle—those dismal things that chemists call the advanced glycation endproducts, the AGEs. These cross-links are nothing but junk that the body has not learned to get rid of. So, said Aubrey, all we have to do is break the cross-links. We can repair the damage of the AGEs. Chemists already know what they're made of. Chemically, they are sugars. All we need is a solvent that can snap off those extra sugars without breaking or fraying the ropes. In other words, we need a solvent that is very highly selective. It has to be selective because most of our proteins, most of our molecular machines, contain cross-links made of sugars. They're not confined to old ropes of collagen. They're everywhere. We want a compound that breaks only the cross-links that have formed by accident without breaking all of the other cross-linked machinery that we need to stay alive. In the ideal situation, said Aubrey, the particular links we need to break would turn out to be the ones that are easiest to break; whereas the links we want to preserve would be harder to break; and with just the right set of chemicals and solvents we will be able to snip the feeble cross-links as fast as they form.

Researchers in big pharma are already at work on this kind of problem. After all, the first antiwrinkling cream that actually erases wrinkles will be worth more money than Viagra. Almost every human hide gets less elastic with age, a condition known as elastosis. Aubrey told me about a chemical that was being promoted at the time by a small biotech called Alteon. Alteon's chemical of interest was purported to break one particular class of cross-links, called dicarbonyl linkages. The interesting thing about the drug was that it was supposed to work catalytically. That is, it broke one cross-link and then was released in its original form to break another.

I followed Aubrey's rap about AGEs with the same polite interest and skepticism that I would have shown a sales rep from Alteon. It's not a bad idea to try to prevent or fix those cross-links. That's a perfectly reasonable goal for medical researchers to work on, since cross-links do so much damage to our aging bodies, inside and out. And, as Aubrey was saying, many researchers have been working on it for years. In fact, I recognized one of them when Aubrey mentioned his name. I'd met him back in the early 1980s, at about the time I met Maria Rudzinska. He was already working on the cross-links problem back then. It was interesting to hear that his ideas were finally being put to the test by a biotech company. I made a mental note: that might be worth checking out. But the specificity problem isn't something to brush away lightly, with wrinkle treatments or anything else. The specificity problem is basic to medicine and always will be. If specificity weren't a problem, you could cure any headache on Earth with a simple surgical tool called a sledgehammer. That procedure would work, but it would have side-effects. If specificity weren't a problem, the historical model for Doctor Faustus wouldn't have gone to jail; his facial-hair remover wouldn't have taken off the patient's face along with the hair. In medicine, you might almost say, specificity *is* the problem. Without it, you've got no one left at the end of the day to pay your fee. "Cure the disease and kill the patient," as Francis Bacon was the first to say. Medicines used to be called "specifics."

Cross-links are one of the simpler items on Aubrey's list of Seven Deadly Things. But as Aubrey laid them out for me, I thought some of his proposals did sound surprising and clever. He told me about his own special interest, the aging of our cells' factories, the mitochondria. He reminded me that because of the low-grade chemical fires that burn day and night in the mitochondria, sparks are always flying around (metaphorically speaking), and some of those sparks singe the mitochondrial DNA. The ancient slaves inside our cells give us our energy, but they burn themselves in the fires of their own furnaces. That damage, the corruption of our mitochondrial DNA, is the second of Aubrey's Seven Deadly Things.

Aubrey told me that he had a way to deal with that damage. The DNA in the nucleus of each human cell contains about twenty thousand genes. But the DNA in our mitochondria is much simpler. It contains only thirty-seven genes and encodes only thirteen proteins.

People who are born with mutations in those thirteen mitochondrial genes are in serious trouble. Because the work of the mitochondria is vital, mutations there can cause rare and horrible diseases of the brain, the heart, the muscles, the liver, the kidneys.

Those precious thirteen are in the hot zone. The DNA that encodes them is subject to as much as one hundred times more cooking than the DNA that is ensconced in relative safety in the nucleus, farther away from the furnaces. If only those thirteen genes were inside the nucleus, they would be safer, Aubrey said. Then they would not be baking day and night in the fury of our molecular furnaces, the mitochondria. According to present theory, in fact, most of the mitochondrial genes have already made just that move. The ancestors of our mitochondria had about a thousand genes. All but those last thirteen genes have migrated to the nucleus.

"So the question is, why haven't these genes moved?" Aubrey asked.

That startled me. I'd never heard the question asked before, or thought to ask it. What Aubrey was describing is, in fact, a fairly common phenomenon: when two living things become closely intertwined, they move genes around to achieve the best, most optimum, fit. Biologists discovered a similar story recently with aphids and the bacteria in their guts. It's a little like the exchanges between cultures—when one tribe conquers and engulfs another (think jazz and blues in America). Gains and losses, as in a marriage. An article in the journal *Science*, describing the shifts of genes from the bacteria to their hosts, the aphids, summed it up neatly: "Any successful relationship demands sacrifices." So Aubrey's question was interesting. If the mitochondria have given up virtually all of their genes and shifted them to the relative safety of the nucleus, then why haven't they shifted the last thirteen?

"There are lots of ideas," said Aubrey. "All but one is complete piffle. The reason is hydrophobicity." For complicated reasons, some of the cell's machinery has to be so constructed that it is hydrophobic: that is, so that its molecules do not like contact with water. The word "hydrophobic" means, literally, afraid of water. And this question of hydrophobicity is vital throughout the cell because most of the fluid inside there, the cytosol, is water. All of the molecular machines that float, swim, hammer, and saw inside the cell have to work underwater. If they are hydrophobic, afraid of water, they may clump and ball up so tightly that they are unable to function. It's the difference between, on the one hand, dropping spaghetti noodles into a pot of boiling water and, on the other hand, pouring in a splash of olive oil. The noodles will swirl around in the water, as long as you stir them once in a while, because pasta is not hydrophobic. But no matter how much you stir, the oil will just clump on the surface, because it is.

Aubrey took a swig from his beer bottle and discovered that it was empty. He said, "Another good thing about drinking all the time is that I keep my voice."

I gave him a look. "Well, *water* will do that, Aubrey."

Aubrey laughed his most charming and disarming laugh.

When he got back from my kitchen with another bottle of beer, he explained a bit more about hydrophobicity. Each and every one of the thirteen genes in the mitochondria encodes molecular machinery that is highly hydrophobic, he said. That may be why those thirteen genes never moved to the nucleus. So, he said, why don't we move them ourselves? We should do what evolution has failed to do and inject good copies of those thirteen genes into the nuclei of human cells, using the procedures of gene therapy.

Here I gave Aubrey another skeptical look. But he was prepared for it. Gene therapists are already able to inject genes into multicellular organisms like flies and mice, and they are growing more capable and sophisticated every year.

"Totally straightforward," Aubrey concluded. "With less than ten million dollars and within five years—or certainly ten years—I could make mice that did not have any mitochondrial DNA."

I asked him how long those genetically-engineered mice would be likely to live.

"No idea," Aubrey said. "If we did that and nothing else, it could be they'd live just a bit longer. But I don't care. It's a candidate mechanism, so let's fix it! If we can fix it, we should. If there's only seven things to fix, then we damn well should. Let's not waste our time arguing that one or two of them might not matter."

That is one of the logical necessities of Aubrey's argument. You need to solve all seven problems at once if you want to extend our lives dramatically. Solving just one or two won't do the trick. We're looking at the sad, familiar truth that if one thing doesn't get you, then another thing will. If you don't get cancer you are likely to die of atherosclerosis. If you don't get atherosclerosis then you are likely to die of Alzheimer's. And so on. To extend human lives indefinitely, to engineer our bodies into a state of perpetual health, we would have to dodge every single one of those diseases, the late-onset diseases. We'd have to figure out how to cure them all or prevent them all, or at least to postpone their onset indefinitely. We'd have to cauterize every head of the Hydra. William James makes this point in another connection in "The Sick Soul," a chapter in *The Varieties of Religious Experience.* "A chain is no stronger than its weakest link," he writes, "and life is after all a chain." In that same chapter, James calls death "the worm at the core" of all human happiness. "Let sanguine healthy-mindedness do its best with its strange power of living in the moment and ignoring and forgetting, still the evil background is really there to be thought of, and the skull will grin in at the banquet."

Then we have that mournful scholar dreaming of his lost Lenore, and the bird that croaks from the bust of Pallas just above his chamber door:

> *"Take thy beak from out my heart, and take thy form*
> *from off my door!"*
> *Quoth the Raven, "Nevermore."*

In any case, it's a simple point. If life is a chain with seven weak links, then you have to fix each and every one of those weak links to strengthen the chain.

Aubrey's suggestion about moving those vulnerable thirteen genes out of the mitochondria was intriguing. I've since talked about it with a number of biologists. All of them thought it was ridiculously complicated and risky, but a few found it interesting, even so. One famous molecular biologist, Seymour Benzer, at Cal Tech, who had taken up the study of mortality in his old age, told me that he and a student had tried to make the repairs that Aubrey was suggesting, in fruit flies, one summer. They ran into a few technical difficulties and he set the experiment aside.

Aubrey went on with his list. First, we have the cross-links that wrinkle our skins and stiffen our veins and arteries and do all kinds of visible and invisible damage to our bodies as we get older. Second, we have the mutations that accumulate in our mitochondria. Third, we have junk that accumulates inside the nerve cells of our brains. Whenever pathologists autopsy the brains of people who have died of Parkinson's, they find Lewy bodies, for instance, which are tiny balls of nasty protein.

These clumps and balls are hydrophobic; so we talked a little more about hydrophobicity, and its importance in the life of the cell. All of our molecular machinery in the cell is made of proteins, and when the cell manufactures proteins, they extrude from the cells' manufacturing sites like long straight noodles of pasta. After these long spaghetti noodles are extruded they fold up almost instantly into complicated and intricate shapes. Their shapes, if they were entered into contests, would win every prize on Earth for architectural, industrial, and sculptural design. It's as if you dropped the noodles into the pot and they did not just cook until they were *al dente*; one of them folded up, in a time much less than the blink of an eye, into a machine that dices, and another into a machine that chops, and another into a machine that blends. And the tiniest differences in these designs can become matters of life and death as we get older. For instance, in some families, people tend to develop Alzheimer's disease very early, in their forties and fifties. They have the bad luck to carry mutations in their genes for beta-amyloid. The mutations make their beta-amyloid more hydrophobic. So it's more likely to clump in their cells. According to present thinking, if beta-amyloid clumps in your skin cells, it may not do much harm. But if it clumps in the nerve cells in your brains, it can do terrible harm, because those cells are so delicate, complicated, and crucial to our functioning as human beings. Michael Hecht, a chemist at Princeton University, is in the middle of a series of experiments in which he inserts various versions of beta-amyloid into bacteria to see if they clump and aggregate. He rigs the experiments so that if the beta-amyloid folds up properly, it lights up and fluoresces a bright green. But if the stuff clumps and aggregates in the cells, it doesn't light up. Again, it's all just simple cooking combined with simple engineering, but at the level of molecules instead of noodles and oil in a pot. Hecht makes random changes in the beta-amyloid and finds that those changes that make it more hydrophobic do make it tend to clump more. The fatal differences are subtle. A basic protein is shaped like a noodle with lots of little attachments called "side chains." If you have all those side chains in the right place, you may live past the age of one hundred with all your wits and memories. But if just one side chain is in the wrong place, your whole family is in danger of developing Alzheimer's in early middle age.

Sitting in my study, Aubrey reviewed the issue of the junk in the brain cells. No one knows how much damage this debris does to the brain and to the life of the mind. No one knows if or how they cause Alzheimer's and other dementias. We really don't know much about dementia, which is not surprising, because we don't know much about how brains produce consciousness. If we knew how the brain makes the mind, it might be easier to figure out why the brain stops making the mind. If we knew how the body makes the mind, we might be able to figure out how a sick body makes a sick mind. Meanwhile the study of Alzheimer's and other dementias is a huge, growing field, and the various schools of thought clash like ignorant armies. Some neurologists think the worst kind of junk in there is the beta-amyloid protein, or BAP; other neurologists blame the tangles, which are made of a protein called tau. These two camps call themselves the Baptists and the Tauists. Battles are fought between the Baptists and the Tauists. It's a small war; but even so, feelings run high.

While Aubrey was telling me his plans to clear away the junk from old brain cells, I heard my wife's steps hurrying up the stairs to my study. She poked her head in the door to tell me some news about a friend of ours. By a strange coincidence, the news had to do with Alzheimer's. Our friend's elderly mother had just been found wandering in a town half an hour from ours. Our friend was at work far away, and she had gotten a call from the police. She needed my wife to go fetch her mother from the station.

After my wife drove off, Aubrey returned to the battles of the Baptists and the Tauists. Each side had its points. "But I don't need to care about that," Aubrey said. "I take the view, no matter what the change is between young and old, if you fix everything, then—"

Just fix every weak link in the chain.

It had taken us a few hours to talk through just three of Aubrey's Seven Deadly Things: cross-links, mitochondrial mutations, and the junk that builds up between nerve cells. Three down, four to go. Aubrey seemed to feel more encouraged than discouraged as he laid all this out. Part of the beauty of his plan in his view was that you didn't need to settle the war between the Baptists and the Tauists, or any other controversy in science and medicine. The thing for us to do is to get rid of all the junk that accumulates in aging bodies. "I just want to fix everything unless I'm completely convinced it's not in the killer camp."

So that famous night before dawn in his motel room in California, Aubrey had scribbled them all down on a sheet of paper, the basic kinds of detritus that accumulate. The list itself was a bit confusing back then. In no particular order, here is one tidy way to sum it up: There's junk inside cells; and there's junk outside cells. There are mutations inside the nucleus; and there are mutations outside the nucleus. There are too few cells; there are too many cells. And there are the cross-links, which stiffen up our working parts everywhere throughout the body at the finest scale. Aubrey had to come up with strategies to fix each one of these Seven Deadly Things. These are the plans that he soon came to call his Strategies for the Engineering of Negligible Senescence, or SENS.

It's a provisional list, of course. Again, the manifold damage we call aging is like the Hydra. If we lop and burn off one head of the monster, the others remain our mortal enemies, and they will bring us down. Most doctors and medical researchers have made their peace with this. They'd be content to solve just a piece of the problem of mortality. If they succeed in treating arthritis or curing Alzheimer's they will slow aging by some small amount. Like inventors and innovators throughout modern history, they will give us the gift of a few more minutes, hours, days, a few years at the most. But immortalists like Aubrey de Grey don't want to slow aging, they want to kill it. To do that, they have to win a war on every front at once. They have to lop off every last head of the Hydra. It would be a labor of Hercules to lop them all off. But we could do it, Aubrey says. And he would be willing to add another to the list if it reared its ugly head.

After half a day of talking with Aubrey, I wasn't sure what to make of him. He did seem enormously well-informed. And he had credentials. He'd hosted an international meeting of gerontologists in Cambridge under the banner of SENS. "They gave me a standing ovation at the end of the meeting," Aubrey told me. "And I'll have to do it again, which suits me fine." And he'd arranged special, smaller meetings of experts to talk about some of his ideas for fixing the Seven Deadly Things.

On the other hand, it all did sound a little crazy. Darwin's mentor, the geologist Charles Lyell, advised him to avoid controversy—it's a terrible waste of time. When you follow the edges and frontiers of science, you try to watch where you step. It's only too easy to waste years in controversy, or step right over the edge. A man with a bottomless bottle of beer, and a beard halfway down to the floor, who claims we can live a thousand years, presents a picture that more or less defines the realms beyond the edge of science, like those sea serpents in the old maps with the legend "Here be dragons."

From my bookshelf, I took down my copy of Bacon's *History of Life and Death*. I read aloud the passage where Bacon explains why we should in theory be able to live forever: "for all things in living creatures are in their youth repaired entirely; nay, they are for a time increased in quantity, bettered in quality." So much so that "the matter of reparation might be eternal, if the manner of reparation did not fail."

I thought Aubrey would agree with Bacon, but he shook his head. "That can no longer be sustained," he said. "It is true if you don't get down into too much microscopic detail. We see no decline in function of tissues until middle age. But the things that cause decline started in conception—or even before, you could argue, in the unfertilized egg. Certainly in prenatal life." Even in the tissues in an embryo, or the cells in a single tissue, slight errors are being made from one reproductive cycle to the next. When cells divide, the changes get passed down. That is one reason that identical twins are never really identical. You could say that junk is already building up in the first moments of the life of the fertilized egg.

"What's going on during early life is a gradual laying down of damage," Aubrey said. "All the same things I've been talking about happen all through life. I'll try to say it concisely," he said, rapping his palms on his thighs. "A forty-year-old is different in composition from a twenty-year-old. In what *way* is that person different? There are no easy answers. The differences are very subtle, very slight. But you know they're significant because the forty-year-old has a life expectancy that's twenty years shorter than the twenty-year-old." Whatever your age, and wherever on Earth you live, your mortality rate doubles every eight years or so, from birth to death. And it doubles because of the buildup of damage and garbage.

Every gerontologist knows about this doubling of mortality rates. This is one way to measure aging: the likelihood of dying at each age. Actuaries call it the "law of mortality." The mortality rate of a man of fifty is many times greater than the mortality rate of a boy at fifteen.

In fact, our mortality rates—over most of the world—double every eight years or so. This is a puzzle: Why should the doubling rate be the same around the world when local populations have such different risks—for instance, low risk of breast cancer in Japan, a tenth what it is in the United States? As a proponent of the theory of the Garbage Catastrophe, Aubrey argues that the rates are so uniform around the world because so many different kinds of junk build up in our bodies wherever we live on the planet.

"So what's going on during early life is a gradual laying down of damage," Aubrey said. We already have the start of atherosclerotic plaques in our major arteries and cross-links in our skin as toddlers. "All the things I've been talking about happen all through life. The only reason it looked to Bacon as just described is that those types of damage, until they reach a threshold, a certain level of abundance—" Until we are thirty or forty, Aubrey said, the damage is too insignificant to matter. "Until then it looks like there's no aging going on."

That really is a sensible description of aging, according to present thinking. Unfortunately, I thought, Aubrey's prescriptions were carefully posed to sound more sensible and plausible than they might to skeptics who are aware of the trade-offs involved. Stimulating the immune system can be dangerous, for instance. The body develops inflammation to try to disperse a foreign body or kill it. And it is usually very effective; but the downside is that cells do it by releasing oxidants, and that's bad. So acute inflammation can be healthy, but chronic inflammation is not. That is why Caleb Finch, of the Andrus Gerontology Center at the University of Southern California, Los Angeles, argues that inflammation may be a crucial problem in aging itself.

But Aubrey had his stump speech about the Seven Deadly Things and he stayed right on that stump. Accumulating damage drives our cells more or less crazy, Aubrey writes in *Ending Aging: The Rejuvenation Breakthroughs That Could Reverse Human Aging in Our Lifetime*, by Aubrey de Grey, Ph.D., with Michael Rae. (On the back jacket, in big capital letters: "PEOPLE ALIVE TODAY COULD LIVE TO BE A THOUSAND YEARS OLD. A LEADING RESEARCHER SKETCHES THE REAL 'FOUNTAIN OF YOUTH.'") The damage, he writes, "forces our cells to flail about in increasingly desperate, disorderly, and panicked attempts to keep their heads above the waters of the aging process." The way to keep the forty-year-old's life expectancy the same as the twenty-year-old's is to keep cleaning up all of that detritus, by stimulating the immune system, etcetera. And we don't have to clean up everything that will ever matter to the aging body; only those insults that matter within our life spans now—only those things that slow us down in threescore years and ten. "Once this is accomplished," Aubrey writes, "our bodies will remain youthful during the years in which they are now undergoing a slow descent into decrepitude." So we will try to stay young and fit while we wait for more help from science, the way other generations strove to stay virtuous while they waited for the Messiah.

Once we did it, once we fixed all seven weak links, eliminated all of Aubrey's Seven Deadly Things, we would live long enough at last to achieve "escape velocity." We would live virtually forever. We would have achieved negligible senescence. At that point human life would be completely transformed, of course. Among other things, virtually everyone on this planet would feel as Aubrey did, that there was little point in having children, because there was so much to do.

Each of us would feel that we had so much life ahead to enjoy just for ourselves.

"We'd have no one under the age of fifty soon enough," he said cheerfully.

I went down to the refrigerator with Aubrey to get him another bottle of beer, and we ran into my two boys. They were fourteen and seventeen years old, and they were curious about him. You don't meet many characters like Aubrey de Grey in small-town Pennsylvania.

Aubrey was wound or overwound, singing his long saga of the Seven Deadly Things, and he went straight back to the top when he saw my boys.

"Suppose we fix aging," Aubrey told them in the kitchen. "So your risk of death is postponed indefinitely. You'd live in the region of a thousand years. *You* have a better chance than *you*, and *you* have a better chance than *you*," Aubrey said, pointing with his right hand to each one of us in order of age, from the youngest to the oldest (me), while he squirreled his left hand deep into his beard.

"But once we have learned to postpone senescence indefinitely, our life span will become limited only by accidents, and that will give us an average life expectancy of one thousand years. So people are likely to live a long, long time," he said. "It seems extremely plausible to me that by then you'd live long enough to live essentially indefinitely."

My boys, both of them science-fiction fans, seemed comfortable with Aubrey's confidence that they would live indefinitely. One of them mentioned *Star Trek* teleportation. "Beam me up." The beam from the spaceship lifts the astronaut from here to there, sometimes thousands of miles away, or more—but maintains the same person.

"Yes," said Aubrey. "That is fast teleportation. This is slow teleportation. You'd be maintaining the same person from century to century by medical means. And if you suffered an accident, eventually we'd know enough to put you back together again no matter what happened."

I protested. We don't know how the brain works. What about the brain, the mind, identity? Aubrey replied that there was no way of knowing what exactly the doctors would have to transfer into the reconstructed brain to make sure that identity is carried over. But in practice he was sure the doctors of the future would be able to do it.

I gave Aubrey another look.

We can't do anything like that now, he conceded. "But it's not implausible for, say, one hundred years from now." To make a map of your patient you'd scan the brain. Then you'd have all the information you'd need to re-create your patient in case of an accident. "Not obvious you could *not* do it." He found such scans perfectly easy to imagine. You'd get one every month. Then if you came to some sticky end, your doctor would use the last scan to reconstruct you. Beam you up. Restore you, and restore your memory files. You wouldn't lose one bit.

Aubrey went on, with the same sort of pleasure with which he'd just been talking about clearing away the junk from aging brains. "Well," he said, "would you really be the same person that went under the truck? I've tried to think subjectively: What is my emotional attachment to the body that went to sleep in O'Hare last night?" He said he was perfectly able to reestablish a sense of continuity after sleep. Why not after a scan? "I think it's very likely."

All this is far in the future, I cautioned my boys. We don't know how to begin to do this now.

Aubrey agreed. But we don't have to worry about any of that today, he said. We are still very attached to our bodies. He used the phrase "meat puppets." We want to keep our meat puppets. If we achieve immortality by uploading our minds into supercomputers, then we will have to say goodbye to our bodies, our meat puppets, forever. "So, not uploading," Aubrey said. "I'll stick with the meat-puppet approach. Of course, if you live a thousand years, driving will be outlawed! It could be a highly risk-averse world." Here he returned to his theme in my car on the way from the airport. If you hope to live a thousand years and you are struck by a cab at twenty-five, you lose an awful lot. "That would piss people off. So there will be an incentive to improve medical care—traditional medical care. And there will be all kinds of safety precautions. Climb a mountain, they'll catch you before you hit the ground if you fall. Automated cosseting. But of course there's only so far you can go. Like how many people you could have sex with without catching something." My boys looked impressed that Aubrey was talking so freely in front of them.

I asked him how long he thought it might be before we arrive at this automated, cosseted world.

"I wouldn't be surprised if it's here in a hundred years," he replied. "I plan to be around. I will warn you that I was surer of that ten years ago than I am now. I feel it's all very well to take this view selfishly. But ultimately if I can do something to add even one day to the human life span . . ."

Here he went into his statistical rap. Already this was the third or fourth time I'd heard it. He was beginning to remind me of a wound-up clock that chimes on the hour, or a salesman who makes the same speeches so often that he forgets what he's just said to you and lives in mortal danger of repeating the same anecdotes two or three times in one pitch. He explained about escape velocity, and saving one hundred thousand lives a day.

While Aubrey talked, I tried to read my boys' faces. No, they did not seem shocked by his confidence that they would live forever. They took their immortality for granted. If anything, I thought they seemed happy to meet an adult who was willing to acknowledge the truth. They told me later on that they thought Aubrey's argument was sensible. He seemed very full of himself, but his premise was only common sense. One of them told me, "I think he is knurd. He is excessively sober." My son had gotten the word from *Sourcery*, a science-fiction novel by Terry Pratchett. "Knurd" is "drunk" spelled backward. Pratchett writes, "Knurdness strips away all illusion, all the comforting pink fog in which people normally spend their lives, and lets them see and think clearly for the first time ever. Then, after they've screamed a bit, they make sure that they never get knurd again."

For his part, when we were back in my study over the garage, Aubrey told me that he found it refreshing to talk about immortality with teenagers. They are people who are positive and adventurous about the future. He feels frustrated when he talks with those who are less adventurous. "That means nearly everybody in influence and power," he says. "Middle-aged and older. They find it so shocking that we might create a world so different from the world they're used to. They're very resistant to even thinking about the desirability of it—that it might be a good thing. People are like that. There's only so much change they can think about. I'm guilty of this myself. Young people talk about uploading. One of your sons brought this up. I just can't see it—can't see it being useful. It seems in no way desirable. But that may be a danger of being over thirty."

Of course, we were getting ahead of ourselves. There were really two enormous questions to discuss: feasibility and desirability. As philosophers say, "can" is not the same as "ought." Aubrey and I agreed that we would save "ought" for another conversation.

I felt sure that the answer to the first question was *no*. The conquest of aging was impossible. The point that bothered me most in Aubrey's spiel was his assumption that we could understand the machinery of our bodies well enough to clean them up. "But we don't have to understand metabolism," he insisted, once again. "I say, go in early enough but also late enough. Early enough to help, but late enough so that you are out of the way of the really complicated stuff."

He saw himself as working in the tradition of the theoretical biologists. "Theoretical biology has an incredibly bad name," he said. "And the reason it's got a bad name is well understood. Since we deal with such complicated systems, biology is a big big subject, and it's very easy if you're an amateur to read a bunch of literature and come up with a nice hypothesis to explain all this data; and if you're careless, you tend to rush into print without checking to see if your idea is consistent with the other 99 percent of data that you haven't got around to reading. This has happened a lot. That's how theoretical biology got into the fix it's in today.

"But the other side of it is that if you have any decent ideas, and the biologists can't see any gaping holes—you do it once and people take you seriously. Twice or three times and you're a phenomenon. So I basically kept my foot out of my mouth for two or three years and everyone was very happy to treat me as a proper scientist, even though I had no idea how to work a pipette." He took a swig of beer and wriggled his fingers together to illustrate his pleasure.

In fact, after his moment of revelation at the Marriott in California, Aubrey had done a huge amount of work with established scientists. He'd kept his job as a computer programmer in the Department of Genetics at the University of Cambridge. It was only in his spare time that he worked on the conquest of immortality, or "the engineering of negligible senescence," the creation of human bodies that hardly age. He was an amateur—but an extraordinary one. He'd published a paper about mitochondrial diseases with one of the world's leading authorities on the subject. Papers with famous gerontologists; venerated epidemiologists; legendary cell biologists.

He was the most accomplished amateur scientist I'd ever met. He was also the most arrogant. "At the moment," Aubrey told me, "probably I'm the only person in the world who has reasonably in-depth knowledge of all the related fields of life extension. That's not going to remain the case for very long. People are going to start putting two and two together. People will start realizing the reason for their pessimism is they haven't been paying enough attention to the *facts*." He railed against gerontologists. "It would be very hard to find anybody to debate me and make a good fight of it without my making a fool of them. Because they are fools. Not in the sense of their intelligence but in terms of what they know. They just haven't done their homework. They're not fools in terms of intellect. But they just haven't had the time or inclination to get the right constellation of knowledge." When Aubrey was explaining one of his most daring and disturbing ideas about longevity, he told me, "In two or three years the whole area will be two or three times bigger than it is now. Due almost entirely to my own efforts."

That evening in Bucks County, my wife and I took Aubrey to a dinner party honoring a friend of ours, a painter. We were celebrating his retrospective at the local art museum. Because Aubrey would not know anyone, we worried that he might feel lost and out of place. He'd had a long day. Besides, I'd gotten the impression from Aubrey's nervous spell in my car on the way to my house from the airport, and from his rapid, thick speech in my study, that he might be shy. A guru needs tremendous force of personality. All in all, I didn't think much of his prospects. At moments as I'd listened to him unspool his spiel I thought he might have something. At other moments, I thought his Seven Deadly Things was nothing more than a list of seven of the hardest problems in medicine. The field of longevity was already full of larger-than-life personalities, dreaming dreams no mortal ever dared to dream before. I doubted that he would find a place at the table of the great world. I was afraid he might have trouble just finding a place at the dinner party.

But I did not know Aubrey. He strode into the party like a conqueror. "Since I've been drinking beer all day I think I'll stick to that," he told our host briskly, when she offered him a glass of wine. Then he seated himself at the center of a long table in our host's living room and took over. Lifting his beer, he began explaining his mission, and the Seven Deadly Things, to our friends up and down the table.

I don't remember every word he said at that dinner. In ten years, the feasibility of his plan would be clear, he said. Within ten years, people would realize that they have been sleepwalking for the last six millennia, "or whatever it is." Soon the explosion of interest in life extension would be a more or less catastrophic phenomenon around the world, instead of the slow steady buildup we're seeing now. Pandemonium! "What will also change is the amount of trouble I'm making," he said. He did have force of personality. He seemed to feed on the stares of our friends. He grew larger and larger in his chair, there at the center of the long table, until he looked like Jesus at the Last Supper. (Since Aubrey predicts the coming of the kingdom of eternal life not in Heaven but right here on Earth, maybe I should call it the First Breakfast.) My wife borrowed my notebook and wrote to me in block letters: "HE IS MORE SURE OF HIMSELF THAN GOD."

After that first meeting, I tried to catch up with Aubrey now and then when he was in the States. When he's in New York on one of his lecture tours he sometimes drops in on Janet Sparrow's lab at Columbia's medical school to see how she is doing with the junk that builds up in the retinas of elderly eyeballs. I joined him there recently. Listening to them talk gave me a glimpse of the different perspectives of a careful specialist like Sparrow and a theatrical figure like Aubrey, who is a general and impresario in the War on Aging. There's a great difference in temperament and tempo between the bench scientists in laboratories, scientists who take things one half-step at a time, and the planners of millennial campaigns. In Aubrey's presence I asked Sparrow what she thought of his idea of attacking and clearing away the lipofuscin from aging retinas.

"Yes, people ask—what about breaking it down?" Sparrow said, speaking very, very carefully. "But then you've got to worry about the health of the cell." Breaking down trash inside living cells might cause new problems, she said.

"We're lucky in the location of the lysosome," Aubrey countered. Because the junk is already packed into the lysosome, the cell's garbage-disposal and recycling unit, it is sequestered from the rest of the cell. "So that sidesteps our ignorance," Aubrey said. Nothing in our lysosome is intended to get out. If through our ignorance we break the junk into toxic by-products, those poisons will still be locked away safely in the lysosomes.

Sparrow did not quite agree. Molecules of lipofuscin do fragment and diffuse out of the lysosomes, she said quietly. Those fragments may be damaging. The garbage disposal is always breaking down and self-repairing, so stuff is always getting out of there and drifting around in the cell, like the dustlike floaters and motes in aging eyes.

In Sparrow's field, as in many specialized areas of medicine, there are debates over whether junk like this does harm or is merely benign, a by-product of the disease process, whatever that may be. For instance, in the study of Alzheimer's, there are those debates between the Baptists and the Tauists. The Baptists think beta-amyloid is what makes us sick and the Tauists think it's tau. Then there are experts who think that neither compound is toxic. They are just innocent by-products. Something else, something that is bad for us, is going wrong in our brain cells. So I asked Sparrow if the same debate applied in the retinas, which are, in fact, derived from our brain cells—our retinas are the only parts of our brains that are not enclosed in our skulls. Are there squabbles about lipofuscin too—with some people arguing that it hurts our eyes and others arguing that it's harmless?

Sparrow explained, very cautiously and carefully, that my question about lipofuscin had indeed been debated for years in her field. Because the lipofuscin in the retina glows in the dark, most specialists now do believe that in macular degeneration, at least part of the problem might be these molecules of lipofuscin. "We're trying to understand if they're negative," said Sparrow. "We think it's increasingly apparent they are."

Aubrey nodded. "The case isn't closed. But it sure ain't gonna do any harm, getting rid of them!" I had a strong feeling that he wasn't speaking to Sparrow or to me. He was speaking to my notebook and pen, and through them to the world. "And if we restore *everything*," he said, "then we're *done*!"

Chapter 8

THE METHUSELAH WARS

"Vermiculate questions," Francis Bacon called them, "fierce with dark keeping."

Controversies—niggly, bookwormy controversies—bedevil scholars in every age. Every field of study has battlegrounds that burn up scholars' time and energy. To the casual bystander it's all academic squabbling and logic-chopping. To the scholars themselves it is almost life-or-death. Aubrey's work is contentious at least in part because it keeps him darting around in the no-man's-land between battlefields. And yet if you take a long view of the Methuselah wars, you can see that they may be winding down, and you can see that Aubrey, or certain key positions of Aubrey's, may just survive.

In the science of life today, the biggest battlefield lies between biologists who study life whole and those who analyze its working parts. They're the "skin-out" people and the "skin-in" people. In the skin-out camp you have the naturalists, the ecologists, the field biologists, the evolutionary biologists. In the skin-in camp you have the cell biologists and the molecular biologists, experts on gadgets and widgets that are too small to see through a microscope.

Those who study nature whole and those who study it at the level of molecules rarely see eye to eye. Skin-out people look at the big picture and skin-in people look at the microscopic or submicroscopic picture. Skin-out people tend to think about the panorama of history, and skin-in people tend to think about the meshing of molecular gears. Francis Crick once scolded Stephen Jay Gould: "The trouble with you evolutionary biologists is that you are always asking 'why' before you understand 'how.'" Meanwhile the evolutionary biologists blame the molecular biologists for asking how and never why.

The evolutionary biologists are tied to Darwin and nineteenth-century natural history. The molecular biologists are tied to Watson and Crick and twentieth-century physics and chemistry. Darwin was the greatest life scientist of the nineteenth century; Watson and Crick made the greatest breakthrough of the twentieth century, and from the beginning they've seen themselves as ringing out the old and ringing in the new. They see the skin-out biologists as describers, anecdotalists, stamp collectors. In universities, molecular biologists get most of the grant money, the new buildings, and the power. They've relegated the evolutionary biologists, ecologists, and naturalists to the corners of the old buildings and the natural history museums. The dustbins.

It has been an epic rift. The skin-in people tend to be excited about the engineering projects they can do. They study the body's works and wonder how much of their new knowledge they can translate into how much power to improve or save human lives. The skin-out people, the evolutionary biologists and ecologists, tend to worry about what we're doing to the rest of the planet, and what we can do to save the other ten or twenty million species that live on it.

In the early years of the molecular revolution, the skin-in people weren't very interested in the problem of aging, because they didn't think there was much they could do about it. One of the young revolutionaries, Leslie Orgel, a collaborator of Francis Crick's, argued that aging is probably caused by damage to DNA. Our DNA is continually jostled and shaken by cosmic radiation from outer space, and by the agitation of the living molecules around it in our own bodies, and these collisions—along with a thousand and one other accidents—can cause mutations. If mutations occur in the egg and sperm cells, they can cause problems for the next generation. If the mutations occur in other cells, they can cause problems for our own bodies, because DNA contains information that is crucial for the cell in all of its manufacturing processes. When the wrong genes get the wrong mutations, you could say that a cell no longer knows how to live. Orgel argued that cells with mutations would make defective molecular machinery, and the defective machinery would then behave badly around the genes, and eventually the cell's production lines would get hopelessly, viciously snarled. Errors would pile on errors. Orgel called this the Error Catastrophe.

Many skin-in biologists found this hypothesis intriguing. After all, DNA is precious. Genes can't be repaired as simply as the rest of the cell's apparatus. When genes are corrupted or destroyed, the cell has lost priceless information. It's bad to lose a cake; it's worse to lose the recipe. If mutations in cells can lead to cancer, maybe they also cause the many kinds of deterioration that we call aging. So the skin-in people were drawn to Orgel's idea; but few of them worked on it. The Error Catastrophe is a difficult hypothesis to test. Each cell contains six feet of DNA, tightly spooled, with about three billion letters of genetic code on the spool. As the cell ages, it gets typos in unique, random places along those six feet of DNA, as well as typos in the elaborate machinery that reads the DNA. How would you keep track of all those typos and prove that they're aging the cell? We all do have days when aging feels like an Error Catastrophe. But it is an ugly hypothesis to study. It would be very messy to prove; and even if you could prove it, what could you do about it?

By and large, then, the skin-in people left the problem to the skin-out people and their arguments about the evolution of aging. To skin-in people, those arguments were not really science at all. If you couldn't understand a problem at the level of molecules, you weren't a biologist; you were just a philosopher.

Then, in the 1980s and 1990s, both camps began to realize that aging might be malleable. Naturally, each camp assumed that the other side's work had to be wrong. But each camp figured out how to make Methuselahs: creatures that live much longer than the rest of their kind.

The molecular work was started by a researcher named Michael R. Klass at the University of Houston. Klass reasoned that at least some of the sources of our longevity have to be in the genes. So he decided to go looking to see if he could find a longevity gene by making mutants in the laboratory and looking for Methuselahs. For his search he used the tiny round nematode worm *Caenorhabditis elegans*. He bred lots of mutant worms by feeding them a toxic compound called ethyl methanesulfonate. Then he grew them in petri dishes, where he let them graze like sheep or cows on lawns of bacteria. Every day he would collect the worms, put them on a nice fresh bacterial lawn, and see how many were still alive. Through the microscope he zoomed in on the worms' throats, one by one, to see if they were still alive, still swallowing bacteria.

Klass created one thousand different strains of mutant worms. Out of that thousand he found just one strain that he considered to be Methuselah mutants. But he noticed that if he put those worms on a little lawn of bacteria in the center of a petri dish, they would wander off the lawn and try to graze on the bare glass. Apparently that strain seemed to have trouble smelling its food. So those worms were hungry. Probably those wandering worms weren't getting enough food, he decided, and that was why they lived a long time. If they lived a long time because they were half-starved, that was not news—they'd be living longer because of calorie restriction. He was looking for a worm that lived longer purely because a mutation had extended its life span. He decided that he had failed. And after doing so much work, Klass concluded, reasonably enough, that aging genes must be very, very rare, if they exist at all. So Klass abandoned the experiment.

Not long afterward, a biologist named Tom Johnson asked Klass if he could look at that last mutant more closely. Johnson looked at the worms through the microscope. He thought they ate fine. Their calorie intake was not restricted. So after a great deal of work Johnson traced the gene that had mutated and made that mutant strain live so long. He named the gene *age-1*. Johnson found that when he raised *age-1* worms in petri dishes at the warm, humid, Floridian temperature of 25 degrees C (77 degrees F), their maximum life span was more than doubled: it increased by 110 percent. A few strains did even better. Their average life spans increased by 120 percent.

At that time, most experts on the biology of aging mistrusted this work. They were evolutionary biologists. They thought it was impossible that a single gene could do very much for life span. According to the line of argument that ran from Darwin to Medawar to Williams to Kirkwood, when bodies age they just fall apart. The disintegration is not programmed. It is not written in our genes. So how could a single gene make so much difference? Something must be wrong with the experiments.

A few years later, a third molecular biologist, Cynthia Kenyon, decided to take up the search for longevity mutants. There was still so little interest in the subject of aging, and the study was thought to be such a backwater, that she had trouble finding a student she could persuade to work on it. When she did, at last, she found a mutant worm that she thought was perfectly beautiful. Under the microscope, ordinary old *C. elegans* worms look granular and ugly, as if they were made of cottage cheese. But these mutants stayed smooth and elegant almost to the end of their lives. Cynthia Kenyon published her first studies of Methuselah mutants in 1993, ten years after the first paper by Klass.

Kenyon's mutants caused a sensation among molecular biologists. The science that grew out of Watson and Crick's "secret of life" had now found the secret of the fountain of youth. If they could make a Methuselah worm, soon they'd be able to genetically engineer a human Methuselah. Whether or not that was true, the link between the gene and the long life of the worms was very clear, news of Kenyon's mutants traveled fast, and here and there other molecular biologists began to enter the field of gerontology. I happened to be present one afternoon at the lab of a grand old man of molecular biology when a young scientist gave a seminar about longevity mutants. Seymour Benzer, one of the founders of the field with Watson and Crick, was fascinated. He seemed astonished to think that we really might be able to understand aging at the level of the genes. He began looking for longevity genes in fruit flies the way Klass, Johnson, and Kenyon had in worms. When he was eighty years old, Benzer discovered a Methuselah fly. He used to talk about that fly in an almost shady way, lowering his voice, making it thin and quiet, as if he and his listeners were convicts in neighboring cells, as if he had to drop his voice and turn it edgewise to slip a scribbled message between prison bars.

The evolutionary biologists remained skeptical about the molecular biologists' Methuselah mutants. To the skin-outs it still seemed impossible that a single gene could matter that much to the life span. Disposable soma theory predicted that there should be many genes that go wrong with age. Aging was a Hydra with at least nine heads. You couldn't kill it with just a single kick. According to disposable soma theory, the Methuselah mutants should not exist. I once talked about them with John Maynard Smith, a grand old man of British evolutionary biology, who worked out some of the mathematical theory that flows from present thinking about aging. He and I met on one of the upper floors of New York's American Museum of Natural History. Maynard Smith was a brilliant scientist who had started out as an aeronautics engineer. He'd helped improve the fighters and bombers of World War Two, the planes that the young airmen of the RAF flew over London and Berlin after burning their initials into the ceiling of the Eagle. After the war, Maynard Smith had done both theoretical and experimental work on the science of longevity. Now he was nearing the end of his life. When I asked him about the biologists who thought it might be possible to engineer a human Methuselah, he shook his head. He said that something seemed to happen when serious people approached the problem of aging and death. They just seemed to go mad.

As a software engineer with a wide range of interests in biology, Aubrey de Grey is in the camp of the molecular types, the genetic engineers, and also in the camp the evolutionary biologists. Still, he's a proud engineer at heart. When I told Aubrey what Maynard Smith had said, he smiled. "John Maynard Smith? I have the greatest respect for his intelligence. But he's an evolutionary biologist. He finds it hard to think in an engineering-type way."

That's the way the two camps flame each other. They fire off those killing salvoes again and again. Once I was talking with a famous biologist—a molecular biologist—who had just joined Rockefeller University, and I asked him if he had ever heard of Maria Rudzinska. He looked blank. I described her work to him, how she had been trying to understand death and dying without looking at genes and molecules, just by watching aging cells through a microscope.

"There used to be a lot of deadwood around this place," he said.

In the 1980s, while the skin-in people were making Methuselah mutants, the skin-out people made their own Methuselahs. They did it the old-fashioned way: not through genetic engineering but through Darwinian breeding experiments. At about the same time that Klass engineered the first Methuselah worm, a young evolutionary biologist named Michael Rose bred Methuselah flies. Typically fruit flies breed at the age of a week and a half. Rose watched carefully and selected only the flies that kept breeding in old age, and he bred those. It was work that could have been done in the nineteenth century just as well as in the twentieth. And it didn't violate the disposable soma theory, because Rose assumed that many, many genes would be involved in making a Methuselah.

It was very hard work. Rose bred millions of fruit flies, and each experiment took between thirty and fifty fly generations. But he found that it was possible to play with a fly's length of days. When he allowed only older flies to breed, generation after generation, they evolved longer life spans; when he allowed only the younger flies to breed, they evolved shorter life spans. Rose and his thesis adviser, Brian Charlesworth, announced the creation of the first of these evolutionary Methuselahs in 1983.

Again, unlike the Methuselahs of the molecular biologists, Rose's were acceptable to the evolutionary biologists. These populations of fruit flies represented the same miracle that has happened again and again in the wild. See, for instance, the bats, the flying squirrels, the flying lemur of the Philippines, all of which are Methuselahs. Any population of living things that finds itself in conditions in which it is able to breed at later ages will begin to evolve longer life spans. Since Rose could replicate that miracle again and again quickly in the laboratory, he argued that adding years of vigorous healthy life must be comprehensible at the level of the genes. He didn't know which genes had changed; evolutionary theory predicted that the change must have been produced by a constellation (or cluster or galaxy) of genes. Evolutionary biologists thought they understood the why of aging, and according to evolutionary theory the existence of a single powerful Methuselah gene was impossible.

Even so, changes in life span could not happen so quickly and so repeatedly, both in the wild and at his laboratory bench, unless it was a relatively simple trick to bring off.

Rose championed the evolutionary Methuselahs and tended to brush aside the molecular Methuselahs. He dismissed those Methuselahs as Johnny-come-latelies. In his memoir, *The Long Tomorrow*, Rose writes, "It is always entertaining when molecular biologists rediscover findings from evolutionary biology. They have such an appealing naiveté, like the moment when my son Darius gleefully discovered at the age of one that gravity would help him knock over a glass of milk."

To the outside world, of course, the battlefields of biologists weren't very interesting. What was interesting was the creation of all these Methuselahs. Cynthia Kenyon, who put the Methuselah worms on the map for molecular biologists in the 1990s, was not only a gifted scientist but young, photogenic, buoyant, and articulate on camera. She was invited on dozens of television shows and news shows to talk about them. When people asked her, Wouldn't it be weird to have adult grandchildren? she would shoot back a snappy answer to the interviewer's questions. "I think every grandparent wants to live to see the grandchildren grow up." She made the study of mutant worms sound sexy. "They are like eighty-year-olds who look forty." And, "I can make your dog live forever!"

Michael Rose also got a lot of press. A generation before, at mid-century, Peter Medawar had dared to imagine that the total span of human life might be lengthened by "stretching out the whole life span symmetrically, as if the seven ages of man were marked out on a piece of rubber and then stretched." Rose declared that ambition too modest. What if we could stretch out the time we spend young without stretching the time we spend withered and decrepit? If we learn to control the genes that govern life span, we could do that. Who knows? We could make youth last threescore and ten years, and age last only one or two years. Certainly we could prolong youth without prolonging age. We could open that door with a few twists of the skeleton key.

These twin victories on either side of the trenches in the Methuselah wars helped revive the field of gerontology. The skin-ins and the skin-outs still don't get along. Skin-out biologists still doubt that you can kill aging with a single gene. Skin-in biologists are still playing with single genes to make more Methuselahs. ("My rule of thumb is to ignore the evolutionary biologists—they're constantly telling you what you can't think," Gary Ruvkun of the Massachusetts General Hospital told a reporter from the *New York Times* not long ago.) Even so, more people on both sides are entering the field, and a resolution of the paradox is beginning to emerge. This battle is beginning to die down.

Evolutionary biologists can see why tinkering with genes might extend the lives of worms, or flies, or mice. It's true that Darwinian evolution does not design bodies to be long-lived. And yet, there are conditions in nature that permit a population of animals to grow up relatively slowly and reproduce at later ages, just as the flies do in Michael Rose's laboratory breeding experiments. Suppose a pair of mice drift out to sea on a log and arrive on an island where there are no cats, hawks, or owls. Suddenly they are safer. Generation after generation, their descendants can flourish if they invest in slow, careful growth, in the kind of cellular quality controls that biologists call "longevity assurance systems." The mice on that island will live long and prosper. Their descendants will inherit their good genes and live longer yet. So aging is in some ways under the control of the genes, even though aging itself is not designed by evolution.

If you look at populations of animals on islands, as Darwin did, you see again and again that when birds or lizards or tortoises get marooned on an island where they are suddenly without predators, they begin to live longer and longer lives. Their descendants are Methuselahs.

This is likewise true of animals that evolve some other kind of protection from their enemies: the shells of tortoises, the wings of birds and bats. All these creatures have evolved adaptations by which they can lift themselves out of the usual rut of danger. With those adaptations they can escape from a thousand ancestral enemies as surely as if they had drifted onto islands. And they too tend to live much longer than sibling species that failed to evolve such a route to safety.

If evolutionary theory is correct about the origin of aging, then life span should tend to lengthen whenever a species escapes a danger that had weighed on it for a long time. Bats make a test case. It turns out that many species of bats are Methuselahs. There are more than a thousand species of bats in the world. They live on every continent but Antarctica and they range in size from the bumblebee bat of Burma, which has a body not much more than an inch long, to the Giant Golden-Crowned Flying Fox of Maitum, Sarangani, in the Philippines, which has a wingspan of five feet. They fit the pattern that present thinking about aging would predict. A greater horseshoe bat weighs about as much as a white-footed mouse, but the mouse lives at most eight years, and the bat lives more than thirty. A big brown bat weighs less than a house mouse, but the house mouse lives at best four years and the bat, nineteen. An Egyptian fruit bat weighs less than half as much as a Norway rat. The rat lives at most five years, while one Egyptian fruit bat is known to have attained the ripe old age of almost twenty-three. The little brown bat, which is the most common bat in the United States, is the size of a little brown mouse; the mouse can live three or four years and the bat as long as thirty-four.

Because they soar so high above their enemies, and can live such a long time, it makes perfect sense in evolutionary terms for bats to invest in expensive maintenance plans—unlike the house mouse or the brown rat, which sprout and die like weeds. It is the same with flying squirrels, flying opossums, and the flying lemur of the Philippines. Strictly speaking, they glide rather than fly, but as gliders they've evolved much longer life spans than mammals of about the same size that can't take to the air. The same principle holds again and again. Naked mole rats are safer than rats and mice because they spend their lives in burrows and tunnels. They can live almost thirty years. There are even parasitic worms that have found their niche in the safety of long-lived human guts. They live a hundred times longer than their cousins in the soil.

Presumably those Methuselahs evolved their long life spans gradually, over many generations. There are also conditions in nature that can induce an individual animal to slow down, grow carefully, and postpone reproduction during its own lifetime. Take calorie restriction. In principle, evolutionary biologists can understand why calorie restriction might lead animals in the lab to slow their rate of aging. It would be adaptive to be able to do that in the wild. During a famine, you don't want to breed; you don't want to bring a new litter of pups into the world to starve. You'd rather slow or suspend your growth, enter something almost like hibernation. You'd want to conserve fuel and energy, riding out the bad times, waiting for better times when it will make sense to reproduce. Calorie restriction probably triggers an adaptation that evolved over many millions of years to help animals cope with drought, famine, and deprivation in the natural world.

Now molecular biologists are finding and exploring some of the mechanisms by which our bodies respond to calorie restriction. In the laboratory, they are studying the genes and cellular tricks that come into play. Many of these genes turn out to be the very same ones that were transformed in the Methuselah mutants.

The quest to find Methuselah mutants has led to a whole bestiary of genes and their products. There is Sir2 (Silent Information Regulator 2), which was discovered in a yeast Methuselah. There is Indy (I'm Not Dead Yet), which was discovered in a fruit fly Methuselah. And on and on: chico, InR, daf-2, fos. Although the field is still tangled and confused, virtually all of these genes seem to be involved in the workings of calorie restriction and the regulation of metabolism. In other words, they connect the work of the skin-outs and the skin-ins; they link the evolutionary theory of aging with the calorie-restriction research of the last sixty years or so.

So far the study of Sir2 has been the most exciting. Sir2 was discovered by the molecular biologist Leonard P. Guarente, at

MIT. Building on that discovery, Guarente and his students and former students began exploring a whole class of proteins called sirtuins (named for Sir2), which are found everywhere in the tree of life, from yeast to mice to people. Work on sirtuins led them to the discovery of resveratrol, which is found in the skins of grapes. Resveratrol switches on sirtuins and prolongs the lives of laboratory mice.

One of Guarente's former students, David Sinclair, now at Harvard Medical School, has helped found a company called Sirtris to exploit the possibilities of resveratrol and find its active ingredients. Sinclair suspects that Sir2 may turn out to have two roles in the cell. First, it works hard to keep the genome stable, to prevent mutations from taking place—a kind of preventive medicine. Second, when DNA does get damaged, it makes repairs: surgical medicine. As our bodies age and we accumulate more and more DNA damage, Sir2 may get so busy doing emergency surgery that it can no longer keep up with its normal, calmer role of preventive medicine. Sinclair thinks that may be the origin of the Error Catastrophe. Although these are still very early days, Sirtris is now testing sirtuin activators in four clinical trials; and Sinclair himself has begun taking daily doses of resveratrol. As he is the first to admit, it is still too soon to say if he is young for his age.

In 2009, a paper published in *Nature* announced another promising drug that slows aging in mammals. This work began with a good idea at the U.S. National Institute on Aging (NIA). The NIH set up the program to allow investigators to test compounds that might intervene in the aging process and extend healthy life span. Scientists anywhere are encouraged to nominate compounds if they can make a case based on our current state of knowledge that they have a chance of making a difference. A series of compounds have now been tested. The first two did not do much for the mice, but the results of the third compound are remarkable.

That compound is rapamycin, an antibiotic that was discovered in microbes found in soil samples from Easter Island. The compound's name is derived from Rapa Nui, which is what Pacific islanders called the place. It targets a piece of cellular machinery that is known simply as TOR (Target of Rapamycin). TOR became a target of interest to gerontologists when work in the laboratory of the molecular biologist Seymour Benzer, at Caltech, linked it to both longevity and caloric restriction. TOR not only helps to shape the life span in flies, worms, and yeast; it is also influential in what is known as the "insulin-like signaling pathway" by which a cell learns if there are nutrients around it.

TOR, like the sirtuins, plays a central role in metabolism. It helps promote the manufacture of proteins; it also inhibits the self-devouring behavior of autophagy. There, TOR seems to be part of an ingenious feedback loop. It enhances autophagy when the cell needs it, and then cranks it down when the housekeeping work is done. When the cell floor gets dusty, it helps draw the broom out of the closet and gets it sweeping. When most of the dust is gone, the broom goes back in the closet. In other words, TOR plays a role in both faces of metabolism: in the creative side, anabolism, and the destructive side, catabolism.

So it made sense to test rapamycin on mammals—on mice.

Testing began simultaneously in three laboratories: the Jackson

Laboratory in Bar Harbor, Maine; the University of Michigan; and the University of Texas Health Science Center. Giving the mice the drug proved to be more complicated than the experimenters expected, and that turned out to be a lucky thing. Near the start of the experiment, when researchers added rapamycin to the mouse pellets, they found that the mice digested it quickly, so that the drug didn't build up to high levels in their bloodstreams. (Human patients have the same problem with rapamycin. They digest most of it in their guts and not much of it gets into circulation. A recent study suggests that taking the drug with grapefruit juice can help.) The researchers were forced to develop a special feed that delivered the antibiotic in capsules for timed release. Developing that special feed took them more than a year. By the time they had it ready the mice in the first cohort of the experiment were already six hundred days old. That put the mice in late middle age. A mouse of six hundred days is about as old as a man of sixty years.

The researchers decided to proceed anyway and the results were more interesting for the delay. Of the female mice in that first cohort, those that did not get the rapamycin had a maximum life span of about 1,100 days. The female mice that got the drug had a maximum life span of about 1,250 days. The maximum life span of male mice was also increased, from about 1,080 days to 1,180 days. If you look at the life expectancy of those middle-aged mice at the time they began to get the drug, the females' life expectancy was raised by 38 percent and the males by 28 percent. (Maximum life span is defined here as the average life span of the longest-lived 10 percent of the cohort. This is a more informative index of maximum life span than the age of the single very oldest mouse in the cohort. In fact, when the researchers analyzed the data, on the first of February 2009, 2 percent of the mice—38 out of 1,901—were still alive.)

We seem to be reaching a kind of hub here. Both the work on calorie restriction and the work on autophagy lead to TOR. And it makes sense that these two lines of research should intersect, because one of the adaptive responses of the body during a famine is to increase the rate of recycling of its own proteins. We start to tear ourselves down faster than we build ourselves up. We get thinner.

Molecular biologists are now studying rapamycin closely and trying to figure out how the experiment worked and why. They want to know why these middle-aged mice did not get thinner on their rapamycin diets. They also want to know whether rapamycin will help to postpone a wide array of late-onset diseases, from cardiovascular and neurological problems to diabetes to cancer. Since rapamycin has serious side effects, they will look for more benign and sophisticated drugs that target TOR, just as they are looking for ever more sophisticated drugs to target sirtuins.

It's intriguing that these new drugs play important roles in pathways that influence so many diseases. With sirtuins, the list includes diabetes, osteoporosis, and cancer, as well as neurodegenerative, cardiovascular, inflammatory, and mitochondrial diseases. With rapamycin, the list is also long, and one particularly promising line of research involves Huntington's disease.

With Huntington's the junk forms because one gene has a sort of stutter in its genetic code. It repeats the letters CAG more than thirty-five times. That unfortunate string of extra letters of code means that the protein is defective; the cell manufactures it with an extra piece or flange sticking out of it and that extra piece seems to make it clump inside the cell.

Recently a team led by David C. Rubinsztein, a biochemist at the University of Cambridge, tried treating these cells in a petri dish by giving them rapamycin, on the theory that boosting the body's ability to take out the garbage in this way might help. It did. Rubinsztein's team also tried rapamycin on a strain of mice that had been engineered as models of Huntington's disease. To test its grip, they let a mouse hold a metal grid with its forelimbs, lifted it by the tail so that its hind limbs were off the grid, and gently pulled backward by the tail until the mouse finally let go. The antibiotic helped sick mice do better on this grip test, and it reduced their tremors.

Most people don't show symptoms of Huntington's until they are at least forty years old, and in almost every case they know the disease runs in the family. These days the mutation is easy to test for. Someday it might be possible to postpone the onset of symptoms and give people more healthy years. In the best scenario, you could delay the onset of Huntington's so long that they would never get the disease because something else would get them first.

The same kind of strategy might work with Parkinson's and other neurodegenerative diseases in which garbage piles up in or around our nerve cells. It may be the same kind of story will be found with the molecular trash known as Lewy bodies, which accumulate in the nerve cells of people who develop Parkinson's, as well as with the trash that piles up in the nerve cells of people with amyotrophic lateral sclerosis (ALS) and other diseases that are rarer and less well known but just as deadly. Typically the damage starts to pile up at least five or ten years before the first symptoms. If the problem could be diagnosed and treated that early, for instance with a drug like rapamycin, which hastened the cells' own brooms, then we might postpone some of the worst diseases of old age, in the best case indefinitely. Two neurologists at Harvard Medical School, Peter T. Lansbury and Hilal Lashuel, note in a review of the problem that this approach has a few strong medical advantages. You don't have to know exactly why the crud is building up and you don't have to know exactly what harm it is doing. All you have to do is encourage the cells' brooms to sweep it up. This is exactly the point that Aubrey de Grey has been making in his arguments about the Seven Deadly Things.

Because most of these nerve cells have to last us our entire lives, they are particularly vulnerable to junk piling up. They can't dilute it by dividing and dividing, like cells in the bone marrow or the gut or the skin. But it is possible that this basic problem of garbage piling up in cells will turn out to be the cause of many diseases of the human body; and early treatment, as here, may turn out to be a way of helping the body stay healthy for longer and longer amounts of time.

Again, the point is that evolution has already given us the broom. Evolution gave us the tools we need for keeping house; evolution gave us the whole house. But evolution did not give us the means to keep house for as long as we would like. Now that we live longer and longer, we wear out the brooms.

The cell's brooms include not only autophagy and lysosomes, but a parallel system involving a molecule called ubiquitin, which tags misfolded proteins in the part of the cell where they are manufactured, the endoplasmic reticulum. Proteins that are misfolded as they come out of the endoplasmic reticulum are carried right back into the body of the cell, into the fluid called the cytosol, where they are dumped into barrel-shaped garbage-disposal units called proteasomes. This particular garbage-disposal process is known as endoplasmic reticulum-associated degradation, which has the acronym ERAD. Here we're getting drawn into the cellular machinery at a very fine and grungy level. The garbage barrel known as the proteasome has a narrow mouth. That limits the size of the junk that can pass into it and the recycled bits that can pass out of it. The autophagosomes that carry trash to the lysosomes are often much smaller than the piles of Huntington's trash they are trying to dispose of. They're like boa constrictors trying to swallow elephants. They may or may not be able to do the job on their own.

God speed the broom. Again, rapamycin has unpleasant side effects when taken long-term. But there may be other drugs that can help the brooms and enhance autophagy. Rubinsztein reported recently that lithium, valproate, and carbamazepine seem to help induce autophagy, too. Combinations of those drugs may do as well as rapamycin with fewer side effects. Of course, as he notes, keeping the housekeeping crews cranked up this way may cause problems of its own. The sorcerers' apprentices may do damage we can't imagine with each extra whisk of the brooms. Or not. Even if autophagy speeds up so much that a brain cell throws out many of its mitochondria, Rubinsztein thinks the cell will still manufacture enough of its energy compounds, its ATP, to function. So you might be able to get the brain cells to stay cleaner and run cleaner with fewer factories and less energy; and the result might be less cellular pollution and longer life.

Huntington's is the disease that first led biologists to the evolutionary view of aging: the view that our bodies are powerless against declines that begin once we have passed the age of reproduction, because evolution is blind to them. That idea was first expressed by J.B.S. Haldane, one of the most brilliant and eccentric British biologists of the twentieth century. Aubrey de Grey likes to quote Haldane's maxim about the acceptance of controversial scientific ideas. There are four stages of acceptance, said Haldane: "One: This is worthless nonsense. Two: This is an interesting, but perverse, point of view. Three: This is true, but quite unimportant. Four: I always said so."

Chapter 9

THE WEAKEST LINK

Toward the end of my summer in London, I went on a day trip with Aubrey. I wanted to visit the site of the very wildest of Aubrey's eureka moments, the place where he had solved the hardest problem of all.

In his program Strategies for the Engineering of Negligible Senescence (SENS), he'd called for mending seven weak links in the chain of life. And in his first flush of enthusiasm for SENS, he'd come up with proposals for mending six of these seven. But even Aubrey had despaired of fixing the seventh link, the weakest link, which is the problem known as cancer. Cancer is caused by mutations in the DNA in the cell's nucleus, and Aubrey didn't see what could be done about that. Our bodies already have huge troops of DNA proofreaders and DNA repairers. The proofreaders evolved long ago to keep cells in multicellular bodies from running amok. They do an extremely good job. But since there are many trillions of cells in a human body, and every one of them is subject to daily mutations, errors do eventually slip through. All you need is one nasty typo in one errant cell among the trillions, and you start building a tumor, and the great chain of your mortal life is broken. A few prophets of regenerative medicine do talk about engineering even better proofreaders than evolution has produced thus far. Aubrey did not believe that this was the answer. He could not see how we could ever do much better at proofreading than nature already does. If we are going to live indefinitely, our proofreaders would have to be absolutely perfect. They would have to catch every single mutation indefinitely, and that seemed like too tall an order.

And then, once a tumor grows it is very hard to kill. Many tumors evolve swiftly because they have escaped the body's proofreaders. They mutate wildly. As Aubrey has written, "Each cell in a tumor is a furnace of inventive potential." Because they evolve so swiftly, these cancers are apt to find ways to resist any attack we mount against them. Some tumors invent ways to eat and digest anticancer medicine. Others invent ways to coat themselves so that the medicine won't get inside them. Then, one day—"one dark spring," as Aubrey writes—the cancer blossoms again. The power of evolution is the secret of the secret of life, the spring of life's creativity. It has brought forth the fruitful tree of life; and in each generation it cuts us down in the most horrible of deaths.

Aubrey was satisfied with the first six proposals in his SENS program. He had demonstrated to his own satisfaction that with work we could fix six of the seven weak links in our mortal chain. For junk inside cells, we could stimulate the cell's garbage-disposal system to do a better cleanup job. In principle, that would cure Parkinson's disease, Alzheimer's disease, and so on. So that was one link mended. For junk outside cells, we could stimulate the body's immune system. That would cure heart disease and prevent strokes.

That's two links. For trouble in the mitochondria, we could inject a healthy set of mitochondrial genes into the cell's nucleus and thus keep aging cells from losing energy and winding down. That's three. For our cross-linked, snarled, tangled proteins, we could find medicines that snip the links. Human bodies would no longer wrinkle or crumple, inside or out. That's four. Some of our cells slow down and become dormant as we grow older, but we can train our immune system to clear those away. That's five. Some cells die, and their corpses pollute their neighborhoods with toxins; the immune system can clean that up, too. That's six.

Those Strategies for Engineering Negligible Senescence were fine and good; but Aubrey could not conquer aging without a cure for cancer. Without that, SENS would do very little to extend human life. Eliminate every other disease of old age and millions of people would live just a few years longer, only to die of cancers of the colon, brain, breast, lung, or skin. Aubrey's first broadside about SENS had left out cancer, but he knew he could not avoid it forever. "If you read that paper closely," Aubrey says, "you'll see I knew damn well the whole business of curing cancer was a massive hole in the scheme." He knew that he had to cure all seven of the deadly ills of aging. The better we do with any six, the worse we will suffer from the seventh. In fact, this trouble is already upon us, in the form of cancer, thanks to the successes of modern medicine. The longer we live, the more likely we are to get it; the younger we die, the more likely we are to escape it. "It's very easy to cure cancer," as Aubrey puts the problem, sarcastically. "All you do is fire all the heart surgeons and so on. It's very cheap as well."

To stay young for centuries, we have to learn to cure every kind of cancer. And we have to eliminate for all eternity, or at least for a thousand years, the chance of developing cancer. Yet the longer we live, the more likely we will be to develop it. As one oncologist has put it, "Advancing age is the most potent of all carcinogens."

"So it all seems a bit sad, really. It all seems a bit of a lost cause," said Aubrey. "How could we ever cure cancer?"

This problem had begun to worry him as soon as the glow faded from his eureka moment in California, the night he'd resolved to fight his "seven deadly things." As Aubrey confesses in his book *Ending Aging*, he feared "that mutations would act as ship-smashing cliffs to any ark that we might build to survive the deluge of metabolism and emerge into an ageless future."

He was so close. His glass was almost full. In principle, he had already cured six of the seven deadly troubles of age.

On your left, the dream of the Phoenix, the bird of immortality. On your right, the nightmare of the Hydra, the metastasizing demon. These are the dreams that have surrounded us from the beginning—life, immortal life; and death, inevitable, inescapable death—and they haunted Aubrey in the starkest possible form.

Aubrey worried about cancer until early in the spring of 2002, when he went to a scientific meeting in Ravenna. He was abstracted during the meeting; in his mind, he kept returning to the problem of cancer. Afterward his hosts organized a tour of the ancient city, and along with the other biologists, Aubrey, still distracted and far away, trooped through the churches, beneath the legendary golden mosaics. He did not know it—he does not care much about history—but

Ravenna has inspired leaps into immortality for thousands of years. Julius Caesar collected his army there before crossing the Rubicon. Dante finished the Divine Comedy there in the last years of his life; the mosaics inspired some of his visions of eternity, like candles lighting candles. Yeats made a pilgrimage to Ravenna. Years afterward, when the poet felt repulsively old, "a tattered coat upon a stick," he remembered the place; he implored all the saints and sages of the past to help him write immortal verse; he saw them standing before him "in God's holy fire / As in the gold mosaic of a wall," and he begged them to gather him "into the artifice of eternity."

After the tour, Aubrey set out for home. He took a bus alone to nearby Forlì and got out in the center of town. Forlì is one of those small Italian towns where you can walk from the bus station to the airstrip. It was a warm day, for March, and when Aubrey had almost reached the airport he stopped at a café to drink and think.

He was sitting alone at his table, when suddenly it dawned on him what to do about cancer. As he lifted his glass of beer, he hit upon what he called a proper cure. He saw a way to fix the weakest link in our mortal chain.

Aubrey thought of the tips of the cell's chromosomes, the telomeres. Everyone in the field of gerontology knows that the telomeres wear away a bit each time a cell divides. According to present thinking, that is why our cells cannot divide indefinitely. We do have an enzyme for repairing the telomeres, called telomerase, but cells run out of it as they age. Then their chromosomes fray, and they come to the end of their tether.

For years, gerontologists have wondered how we could supply our aging cells with more telomerase, and live longer. At the same time, many cancer researchers have wondered about the opposite problem. They would like to find ways to eliminate telomerase from tumors, so that cancer cells would cease to multiply. Cancer cells carry mutations that allow them to make plentiful supplies of telomerase, and that is one of the reasons they have become immortal.

In his first thoughts about SENS, Aubrey too had hoped to eliminate telomerase from cancer cells. But there is a problem with that approach, as there is with every attack on tumors. As long as our bodies carry the gene for telomerase, a cancer cell can always find a way to make more of it. Somewhere in the course of the innumerable random mutations that take place in our trillions of cells, there will always be one rebel that chances upon the trick, makes itself immortal, and multiplies out of control.

Aubrey's insight was this. If we were to eliminate just one gene from the body—the gene for telomerase—then every cell in the body would be unable to repair its telomeres. No renegade cell could rediscover and re-create telomerase. "Creation of a new gene out of nothing does of course occur on evolutionary timescales," as Aubrey has written; "but that takes many, many generations." Even if our bodies lasted thousands of years, we would not have enough cells or enough time to achieve it. The secret of regeneration would be lost. No cell could ever build a tumor, because it could not divide often enough to get out of control. Eliminate that single gene from a human being, and not even our stem cells would have telomerase. Stem cells normally have plenty of it; they need it to replenish our bodies with young cells as our old cells wear out. Without telomerase, even stem cells would reach their limit early. That way they could not run wild with cancer. In a sense, every cell in the human body would be sterilized.

We could maintain tissues like blood, gut, and skin by periodically reintroducing new stem cells. None of *them* would have telomerase, either. We would reseed the body with them, and repeat the operation again ten years later. "Then we'd have to do it again, indefinitely," Aubrey says. "But the point is, we'd never introduce a cell that had the capacity to mutate into a cancer."

That was his eureka in Forlì: take the telomerase gene out of the body!

In many ways this is a desolating vision, a disastrous idea, because without telomerase the body could no longer regenerate itself in the places that need regeneration most. Our skin and the lining of our gut, our outer and inner linings, are always repairing and replacing themselves because they get the most wear. At regular intervals we would introduce stem cells into the body to rebuild those outer and inner linings. We would insert stem cells that we had genetically engineered, cells with abnormally long telomeres. And before those stem cells began to wind down, we would just top up our tissues with more of them. That might not be an easy procedure. No one knows how to do it now. But eventually it would be just one more medical routine for the young, healthy immortals of our boundless future. When cells escape into cancer, they become immortal. We would prevent the birth of those cells and become immortal ourselves.

Aubrey calls this plan, his cure for the "seventh deadly thing," Whole-Body Interdiction of Lengthening of Telomeres (WILT). It is an ugly idea, as Aubrey himself is the first to admit. Not everyone will find it attractive. He writes, "The idea of eliminating from the body a function known to be essential for survival is a conceptual leap that takes substantial justification even to contemplate, let alone implement." He believes most of us will not see its appeal until medicine has managed to cure the other diseases of old age—until we have figured out how to prevent heart attacks, strokes, Alzheimer's disease, Parkinson's disease, and diabetes. At that point, however, so many people will be living long enough to get cancer that we will be willing to undergo something even this traumatic.

In the WILT procedure, patients would undergo periodic bouts of chemotherapy to kill all the cells in the bone marrow. Then they would receive injections of bone marrow in which the cells had no telomerase. Aubrey estimates that they might need new bone marrow transplants every ten years or so. They would need replacements of their skin stem cells every ten years or so, too. The same is true of the innermost layer of the lung; but "there is no reason to think that we won't make quick and relatively painless progress on this front once we put our mind to it," Aubrey writes in *Ending Aging*. Getting fresh stem cells into the gut might be more difficult but could be accomplished by the same general tools and techniques that people endure when they get a colonoscopy.

By adding fresh stem cells wherever and whenever they were needed, we would keep reseeding the body.

Of course, we would have to denude the body of native stem cells first. That would require high doses of chemotherapy. The treatment would get rid of the cancer and clear the body of all its fertility, which we would then continue under new management. It would be a far more brutal treatment than a radical mastectomy. But, then, think of the painful and expensive procedures that many people are willing to go through just to look young, Aubrey argues;

many people pay for chemical or laser "peels" of skin even if the benefit is merely cosmetic. His denuding of native stem cells would allow people to continue to *be* young, not just look young. Since we have no way of curing cancer now, and can't imagine a way that would be foolproof, or evolution-proof, and since the better we do at living longer, the more cancer will attack us, Aubrey thinks we should start working on WILT.

Might cancer cells find some surprising way to escape even this attack? Cells seem to be able to lengthen their telomeres with enzymes other than telomerase. But if that happens, Aubrey says, we will figure out what those enzymes are, and delete them, too.

The transformation of the human body to a totally WILTed state would have to be performed gradually. We would have to endure the rigors and horrors of all that chemo. Men might become permanently sterile and might want to set aside sperm first, to be frozen and stored in fertility clinics. But our risk of cancer would no longer rise with age; it would actually decline.

Not long after he got home from Italy, Aubrey convened a panel of experts to consider WILT. One of the participants in his WILT summit, Nicola Royle, a senior lecturer in the Department of Genetics at the University of Leicester, refused to have her name attached to the paper that resulted. But Aubrey puts a positive spin on that. It wasn't that Royle didn't think his idea would work. She was bothered only by the goal itself, the creation of nearly immortal human beings.

Aubrey had now gone from his starting point to the very limit. He'd begun by imagining that we might keep aging bodies alive by clearing away debris. With WILT, he was envisioning an overhaul of the body. To do what he was describing would be to remove from the body its own powers of rejuvenation and to assume this power and responsibility, entirely and permanently, for ourselves.

"Now, it's critical to remember, we're talking about a *proper cure*," Aubrey said. "Not just to postpone cancer by ten years. We've got lucky here, with evolution having given us this window of simplicity in the middle of a highly complicated cause of events. Cancer is compositionally simple and we have that window to get rid of it." All we have to do is kill one gene—the gene for telomerase.

People have never thought in these terms before, he said, because people have always thought of the body as requiring its own powers of rejuvenation. "The epidermis does constantly renew," said Aubrey. "And it renews from stem cells at the bottom. And those cells do express this gene telomerase. If they didn't have it they'd conk out, and we'd end with no skin. The same is true of the blood, and the same is true of the gut. The same is true of quite a lot of tissues that we rather rely on.

"So this seems like a bit of a showstopper on the face of it." But really, again, all it would take is the subtraction of one gene.

"Why cure cancer *that well*?" Aubrey asked me, when he first laid out his vision of WILT. "Because if you can cure cancer—I mean really cure it—then you've actually done the hardest thing there is in curing aging."

And how soon did he think we could take over the body's powers of rejuvenation?

"It could take ten years; it could take twenty years; but it's not a century away."

* * *

From the moment Aubrey told me about WILT, I knew that I would have to see the place where he had his vision. I thought it would make a wonderful story that he had found his path toward eternal life after wandering in Ravenna. For his part, Aubrey was perfectly happy to lead me to the place where he'd had his eureka moment. He was pleased that I was willing to take WILT so seriously, because most of his colleagues in gerontology thought the idea was crazy. In fact, they thought WILT was by far the weakest link in his scheme. Their objections were numerous. How would you eliminate the gene? How would you deal with the side effects? How would you carry out the necessary procedures: reseeding the bone marrow, the gut, the skin, the lungs? It would be far, far worse than conventional chemotherapy. *Cure the disease and kill the patient.* Biologists who hear Aubrey's idea for the first time often become furious. "WILT is clearly nonsense and the main reason why so few scientists take him seriously," says Jan Vijg, the cancer specialist and gerontologist at the Albert Einstein College of Medicine, who is one of Aubrey's strongest supporters among established gerontologists. "This has nothing to do with disliking Aubrey or seeing him as a competitor or whatever. WILT is just sheer nonsense."

But this is the point at which Aubrey feels that conventional scientists reveal their fatal lack of imagination. What they don't understand, what they don't factor in, is the way medicine will begin to accelerate once we achieve our first modest successes in the war against aging. Once we realize that Aubrey is right in broad principle, and aging can be cured, there will be no stopping us. There will be no obstacle we can't leap over. That is why he has no patience with the pessimism of the gerontologists or the modest optimism of the demographers. When demographers say that during the next century we may gain another decade or two or three in life expectancy, they merely extrapolate from the history of human health in the nineteenth and twentieth centuries. Aubrey calls them "extrapoholics." If we move fast enough, with each researcher building on the work of the immediately preceding researchers, then we will achieve what Aubrey calls escape velocity. Escape velocity, he's written, is "the point when improvements to the comprehensiveness and safety of human life-extension treatments are being made faster than people are aging: that is, when the remaining average life span of those who are receiving the latest therapies, and who are of the age that derives the most benefit from those therapies, begins to increase with time even though they are getting chronologically older." In other words, we achieve escape velocity when science is adding to our life expectancy faster than we are living it—or, in Aubrey's metaphor, when the engines of biomedicine are lifting us upward faster than the forces of decline and decay are dragging us down.

"Escape velocity," Aubrey says, with satisfaction: "it's a bit of a glib phrase but I think it does the job as well as any other pair of words."

His faith in the coming of a new millennium pushes Aubrey even farther out into the dark—or into the boggy fringes of his field, where the footing is treacherous. It puts him at odds with all the conventional gerontologists who talk about adding just a few more comfortable years to our lives. This conservatism of theirs is what he rails against, above the babble of voices in the Eagle. Most gerontologists are so timid! They're happy to trumpet such a modest research program. They're happy to agree with the assumption that we can't live forever. "Most other people are using this as a *funding strategy*!" he cries indignantly. "So it means that we have *repulsively* political phrases like, 'Our goal is adding life to years, rather than years to life.' I mean, I throw up when I hear that! I have no words to describe my disgust for that." His delivery is astonishingly quick, as if he were dashing madly upstream with water sheathing and coating the rocks. "They think it's going to be what politicians want to hear—what purse-string holders want to hear. And that *is* what they want to hear. But the fact is, it's a *lie*!"

And he gives his listeners a cosmic look that says: The victory is infinitely great and just ahead. Follow me!

So, early one summer morning I took a train from London, and Aubrey took a train from Cambridge, and we met at Stansted Airport for a flight to Forlì.

The ticket line was moving slowly, and while we waited I studied the posters overhead. In an ad for Luxury Hilton holidays, a young woman stands at the beach in a red bikini—laughing. And in an ad for Vodaphone, two young men stand at the beach, laughing, with two young women on their backs, also laughing. The women are talking to each other on their cell phones. And the caption is: "How are you?" Another ad, aimed at a more staid crowd, is for *Science and Health* by Mary Baker Eddy: "Fuel for your Spiritual Journey." There is a testimonial from a middle-aged man, L. Rodriguez, business entrepreneur: "I felt so unsure about the future . . . until I read this book."

Aubrey had arrived at the airport before me, and he'd already bought his ticket. He looked a bit testy as he waited for me to get mine. "The trouble is, you're costing me valuable drinking time," he said at last. We arranged to meet in the airport bar.

When I found him again, the sight of him there at the bar gave me pause. Most of the time, I marveled at and was amused by his drinking. He carried it off with so much dash that I rarely questioned it. When Aubrey first told me about his work in Pennsylvania we'd talked for a day and a half. On the second day there was already a powerful yeasty smell in the room from the day before. My study smelled as sour as an old pub. I couldn't help counting when I cleaned up: eighteen bottles of beer.

Now Aubrey glanced up from his little round table at the airport bar and smiled. His mood had improved. "It's days like this that I feel particularly gratified that I don't need breakfast," he said.

On the plane to Forlì, Aubrey showed me the latest issue of *Fortune*. It included a profile of Aubrey, with photos. "The *Fortune* photographer had a business card with a list of the famous people he's photographed," he said cheerfully. "So my ambition is to get on that list."

We landed in Forlì and set off to find the bar where he'd had his idea. To pass the time as we walked through the streets, Aubrey asked me riddles. How many three-letter words could I think of for parts of the human body? The sun was so much brighter in Italy than in England that the black asphalt looked white. A woman on a motorbike whizzed past us, smiling and saying into a cell phone: "*Pronto. Pronto.*"—"Hello. Hello." We passed alleys and shuttered stone buildings that looked blind. Pigeons in a dry fountain, on the corso della Repubblica. The long lane was like the long straight shot from birth to death that people can see from here, in the old Italian sun. The churches and stone walls seemed to be exerting gravity, as if they would pull you down, as certainly as falling. Apparently Aubrey was not susceptible to these suggestions. The novelist Shirley Hazzard has written, "In Italy we learn . . . that the ability to rise to the moment, to the human occasion, is linked to a sense of mortality"; but in Italy Aubrey had drawn the opposite conclusion, that we might escape from death's gravitational pull once and for all.

The scene of Aubrey's vision, when we found it, was nondescript: just a coffee shop with a few scrappy tables outside. Potted plants in concrete marked out a small space on the asphalt for the tables. But the place was closed for the month of August. The windows of the shop were papered up with local newspaper. A cupboard the owners were throwing out or trying to give away stood by the door. Aubrey leaned against it in the shade. An old man passed on a bicycle. Across the street there was a newsstand, and a joint called Blue's Bar. The life-giving, carcinogenic sunshine was intense now: steep noon Mediterranean sun. Roosters crowed from an overgrown backyard.

Aubrey explained that he'd sat outside that day, facing the street. He'd ordered a Tuborg. It came in a very large bottle—which must have been a liter bottle. "A young woman served me. I can say '*Birra*.'" There was only one type of beer. The place was basically a coffee shop. "It was pretty deserted—I was the only person here. The temperature was warm, but not uncomfortable. I was grateful for the beer.

"I was on my second one when I had my critical idea. I didn't need another one after that, because I knew it was an important idea, and I was pretty happy. I just exclaimed to myself, then got up and walked in a jaunty manner in the direction of the airport."

I told Aubrey I found it interesting that he had arrived at his secular vision of paradise in the place where Dante wrote his own—a fabled place in the history of human yearnings toward immortality. But to Aubrey, the setting seemed to be a matter of complete indifference. Nothing in Tuscany seemed to have impressed him on his passage through the first time, either. When I asked him what he remembered, he said, "It was really a rather uneventful meeting. Perhaps that's what cleared my brain afterward."

At my suggestion, we had planned to visit some of the great mosaic-lined churches of Ravenna after our stop in Forlì. We found a bus in the center of town. I pressed Aubrey again about events surrounding the meeting. He did vaguely remember an opera singer. "I just could not believe so much noise could come out of two lungs! Obviously impossible. And she was a little girl as well. Adelaide's size, if that." When I pressed him some more, he grew impatient. "There were a couple of interesting places, certainly. Old palaces."

He doesn't like to travel; he doesn't like to eat; he doesn't have time for anything but drink and work, although his immortality project forces him to trot around the globe. He looked lanky, even skeletal, loping along through the streets of Ravenna. He got a lot of long, frank, solemn stares from children under five, but he did not seem to notice. "That's right," he said, in one square, "we ate most of our meals here. I'd completely forgotten that till I saw that repulsive tablecloth."

Aubrey isn't particularly interested in children himself. In any case, there are fewer children on the streets in Italy and in much of Europe than there used to be. As people around the world live longer, many of them decide to have fewer children. At the turn of the third millennium, seventeen countries in Europe recorded more deaths than births, notes the demographer Paul Demeny: Belarus, Bulgaria, Croatia, the Czech Republic, Estonia, Germany, Greece, Hungary, Italy, Latvia, Lithuania, Moldova, Romania, Russia, Slovenia, Sweden, and Ukraine. Around the world, birthrates have dropped from six per family in 1972 to a bit less than half that now, and some of the lowest birthrates anywhere on Earth are to be found in the cities of Italy. In some Italian towns, the rate is less than one.

"A lot of people point out overpopulation," Aubrey likes to say, when anyone brings up that objection to his immortality project. This will matter in principle, he says—but we woudn't have a big problem with too many children for a hundred years. In fact, people might wise up and not bother with them for an indefinite number of years. "Another way of looking at it—who cares? This is the way I like to look at it," Aubrey says. "We've got a chance of saving people's lives—and we have to do that. Letting people die is bad in the same way killing people is bad. So we've got to do it. Even if we needed severe birth-control measures. So we *have* to do it. And people are very shocked when I say that. Especially when it goes further—when it means we'll end up in a world with more or less no children. Get over it! And people are not happy with that. But I don't know. Maybe it's just my personality, but I prefer to cut the crap and get to the chase."

We walked to Ravenna's Museo Nazionale. The cobblestones

were gritty with gravel, and the marble steps were worn as smooth as stones on a beach. "My mother's so good to me," Aubrey said, stopping at the museum gift shop. She didn't ask him to visit often. "Her one condition is that I send her a postcard from wherever I go. And I'm enormously religious about it."

I flipped over his postcard and read the caption aloud to him: "Vault and Lunette with Good Shepherd."

"I don't spend much time choosing which postcard to send," said Aubrey.

Headless marble torsos loomed over us, attached to marble pedestals by hooks of steel, and then a life-size Christ, crucified: a great wooden Y. I asked Aubrey if he had gone to church as a boy. He said his mother had raised him as an Anglican. "She used to send me to church once a month. She gradually started going less and less. By the time I was ten, we went just at Christmas and Easter." At the time of his confirmation, he was a student at Harrow, a school in northwest London that has been educating boys since the year 1243. "It was a complete nonevent as I recall. I don't go as an adult. Churches are emptying out, to a large extent."

I asked him if he ever wondered why that might be.

"Not my area of expertise, your honor." He snorted and blew air out through his lips.

The courtyard of the Museo Nazionale looked a little like courtyards at the University of Cambridge. "We don't have so many fragments there," said Aubrey. "Then again, our things are only about one-third as old."

Ravenna was the capital of the Western Roman Empire. It was also the capital of the Kingdom of the Ostrogoths.

And when the Byzantine emperor Justinian retook a piece of the boot of Italy, he kept Ravenna as his capital. As a port on the Adriatic Sea, it was a convenient place from which to sail to Byzantium. During Justinian's reign many of the famous mosaics were made, enclosed in plain brick churches, like souls inside bodies. Yeats called the mosaics "monuments of unaging intellect" in his great late poem "Sailing to Byzantium."

"You realize that if ages stretch as you predict, then these antiquities become meaningless," I said.

"That is my ambition," said Aubrey. "I look forward to that." He discreetly checked his watch.

Aubrey and I wandered, to his growing disgust, through the churches of Ravenna. In the Tomb of Galla Placidia, stags (representing souls) drank from a mystic fountain encircled by curlicues of greenery, as if all life had turned into music. Dante's heart would have been touched by these scenes as he finished his *Divine Comedy*, in exile from Florence, during the last four years of his life. His epitaph concludes: "Here I lie buried, Dante, exile from my birthplace, a son of Florence, that loveless mother."

At the Basilica of San Vitale I pointed out the peacocks that face each other on a stone sarcophagus. There were rows of them, symbols of immortality, waiting for resurrection. And then there were still more symbols of eternity and immortality: mosaic birds at a mosaic fountain. The church was built when the Goths were still in Rome. Among the pillars and the groin vaults of the chancel, the Lamb of God was framed in a wreath against a night sky full of gold and silver stars.

"Well, they're fixated, really, in this place, aren't they," said Aubrey, with a donnish drawl. "Can't imagine *why*." That killing drawl—that slight rasp of the blade. I asked him if he'd learned it in college.

"It's sort of more Harrow than Cambridge," Aubrey said.

"It's like the drop shot in tennis," I said.

"A part of my national heritage."

While I admired San Vitale, he settled himself in a pew to wait for me. A chair fell over somewhere in the church with an echoing crash. Then the organ began with a portentous chord, another crash. Aubrey glanced at me with an eyebrow drolly cocked. The chords of the organ went rolling like the waves that washed the marbles smooth—groaning and building toward some great convulsion to which fewer and fewer would aspire. Aubrey had bowed his head as if in prayer. When I offered him a penny for his thoughts, he said he was hatching plans to raise the circulation rate of his journal, *Rejuvenation Research*.

As we walked on from church to church through the streets of Ravenna, facing into our own lengthening shadows, Aubrey diverted himself by wrapping up the riddles of the day. In Wonderland, the Mad Hatter asks Alice a riddle without an answer. "Why is a raven like a writing-desk?" But in Aubrey's world, puzzles are made to be answered, and I should answer the questions he'd put to me that morning: How many three-letter words for parts of the human body? (Answer: nine—arm, leg, eye, ear, jaw, gut, toe, lip, and hip.) Now, in the sinking sun, the failing light, he just wanted to go home.

I'd hoped to find Dante's tomb before we left, but Aubrey was getting very tired and sulky. "Maybe, I think, we've seen it," he said. "They all look the same after a while."

Our last stop was the Baptistery of the Arians ("sometimes called the Neonian Baptistery in honor of Bishop Neon who had it decorated during the middle of the fifth century"). Aubrey remembered it vaguely—he'd stayed next to it. According to my guidebook, it was probably an old Roman bath. The Nymphaeum was decorated with mosaics at the order of Bishop Neon sometime after 452. Directly above, on the ceiling, Jesus was baptized in the Jordan, with a dove descending. Aubrey took a chair with a little groan of pleasure and a smothered yawn. The other tourists around us craned their heads up, and he let his own head fall back, too.

The ceiling was said to have inspired Dante's vision of Paradise. Jesus Christ is baptized at the very top of the dome, in the medallion in the center. All the mosaics and the architectural elements are arranged beautifully to lift your eyes toward that central medallion. The dome is circular. The interior includes apses, arches, columns, windows, niches, porticoes, spandrels, mosaics of thrones and altars. The spandrels are decorated with mosaics of the Prophets and tendrils of acanthus. In the scene within the medallion, against a background of gold, there is a rocky riverbank with radiant flowers, the blue water of the Jordan. A river god holds a green towel to dry off Jesus. The river god has green hair and beard, and a green staff, along with the green beach towel.

High up in the dome, Saint Peter and Saint Paul lead the Apostles, dressed in gold and silver tunics, in solemn procession. They seem to go around and around like one of the "eternal wheels" that Dante saw in the dome of heaven, and the medallion seems to spin like a cosmic pinwheel.

Below the medallion, in his chair, Aubrey looked almost martyred. His face was pale. His cheeks were hollow. His beard hung a good distance down his chest. It was a better beard than John the Baptist's; longer than the beards of the Apostles; much longer than the beard of the young Moses on his hike up Mount Sinai, where he pauses to relace his sandal.

Well, why would Aubrey be moved by any of these saints and sages in their holy fire? Aubrey has his own hopes. We are hurtling toward a sort of technological supernova, an intelligence explosion, a Singularity. The Singularity will bring a golden age. Not long ago, he wrote an online paean to the Singularity in which he concluded, "Humanity will at that point be in a state of complete satisfaction with its condition: complete identity with its deepest goals. Human nature will at last be revealed."

To Aubrey, the failure of our collective will, of our human nerve, is the greatest obstacle to the achievement of escape velocity. Our blindness to what we can be is what prevents us from moving toward the Singularity. We are the weakest link.

The pilgrims and tourists in the church made their way around him where he sat. They glanced at him, at his pallor and hollow cheeks, and they averted their eyes as if here must be a man who was more serious about immortality than they.

The swirling gold world in the mosaics around them suggested grace, and chaotic snake rings of gold against the black—grace unfolding against the blackness of space or the intense inanity of nothingness, not-being. In his weariness, Aubrey looked like one of the saints or hermits come to life—with no time or patience whatever for the world he had just come out of, interested only in the world to which he was going, or hoped to go, the route to the next world. The past held no interest for him, and the present world interested him only as a portal to the next. In that way, at least, he was not unlike the saints and martyrs who regard him just as stiffly from above—returning the gaze of oblivion.

There Aubrey sat with his martyred look, dark, hollow-eyed, and grave—the mark of thought on the pallor of the face. Under the great mosaic dome of the Battistero Neoniano, his head was thrown way back. His eyes were closed, his hands folded.

The pilgrims and tourists tried not to stare at him.

A baby in a stroller gazed at the beard, solemnly fascinated. The parents politely hurried the stroller on by.

But Aubrey David Nicholas Jasper de Grey was fast asleep.

PART III

THE GOOD LIFE

So teach us to number our days,
That we may apply our hearts unto wisdom.

—PSALM 90

Chapter 10

LONG FOR THIS WORLD

Mortality is at our core. We are long for this world, compared with life in the microcosm of the paramecium, the bacterium, or the *Tokophrya* standing on the pillar of its holdfast. We have a greater portion of time than most of the other living things with which we share this planet. And yet how we long for this world, how we wish we had more years to explore and enjoy it! How sharply we feel, at every moment of our lives, that mortality is deeply ingrained within us!

"To be a philosopher is to learn how to die," said Montaigne. But as a thinker during the Renaissance, he didn't have much time to learn to do it. He wrote in his tower, in his final essay, "Experience," "I have recently passed six years beyond the age of fifty, which some nations, not without cause, have prescribed as such a proper limit of life that they allowed no one to exceed it. Yet I still have flashes of recovery."

I'm glad we live at a time when a writer who has just turned fifty-six is not all that old. (Flashes of decrepitude here, but still young enough to go forward, I hope.)

"Write as if you were dying," Annie Dillard advises in her book *The Writing Life*. "At the same time, assume you write for an audience consisting solely of terminal patients. That is, after all, the case."

And that has always been the case, although now we live in a moment when, as we philosophize and grope toward wisdom, we can wonder just what and how different the term and the sentence may be, and if, and when, and what then.

From the beginning our philosophers have tried to teach us how to die, and our poets have taught us that to contemplate death is to learn how to live. Seneca wrote, "We must make ready for death before we make ready for life." "You have noticed that everything an Indian does is in a circle, and that is because the Power of the World always works in circles, and everything tries to be round," said Black Elk, the Oglala Sioux holy man. "The sky is round, and I have heard that the earth is round like a ball, and so are all the stars. The wind, in its greatest power, whirls; birds make their nests in circles, for theirs is the same religion as ours. . . . Even the seasons form a great circle in their changing, and always come back again to where they were. The life of a man is a circle from childhood to childhood, and so it is in everything where power moves." Walt Whitman ends his poem "Out of the Cradle Endlessly Rocking" with a paean to "death, death, death, death."

I once met Aubrey and Adelaide at the Eagle with my family; he'd offered to take us punting on the Cam. But it rained that day, and we ducked into the Fitzwilliam Museum instead. The glass displays in the museum included one of Isaac Newton's notebooks and a few of Charles Darwin's letters. Aubrey strode through the halls of treasures at the same clip and with the same degree of interest with which he'd hurried through Ravenna. He was trying, as always, to recruit my boys to his cause. At one point I stopped at a glass case to read the manuscript of John Keats's "Ode to a Nightingale," which the poet wrote one morning on Hampstead Heath, a short walk from the house where we were staying in London.

I have been half in love with easeful death. . . .

Keats had a year and some months to live when he wrote that line, at the age of twenty-three. He had already lost his brother Tom to tuberculosis, and caught the disease himself. While I leaned over the glass museum case, one of my sons squatted on the floor with his back against the wall, and Aubrey settled down right next to him, urgently explaining his own plans for the engineering of thousand-year lives.

From the first age to our own, mortality has been the theme of writers, including the writers who loomed like immortals to my generation, the giants whose very names can still make us feel as small and hopeless as epigones, even though they are all going now or gone, after all that jockeying for immortality. Norman Mailer wrote about WASPS: "They had divorced themselves from odor in order to dominate time, and thereby see if they were able to deliver themselves from death." Saul Bellow took John Cheever to the Russian baths in Chicago. "Wreathed in vapor he looked more immortal than I," Cheever reported in a letter to his brother, "but I think he was trying." "God save us from ever ending, though billions have," John Updike wrote in his last cycle of poems, *Endpoint*, when he was dying of cancer at Massachusetts General Hospital.

Mortality, impermanence, ephemerality: this has been the great theme of modern science, too. Galileo's discovery of sunspots ran counter to traditional astronomy and its view of the sun as immortal. People had always thought the sun was perfect, eternal, and spotless; he argued that the sun could be mortal and decay—like the rest of us. "It proves nothing to say . . . that it is unbelievable for the dark spots to exist in the sun because the sun is a most lucid body," Galileo wrote impatiently. "So long as men were in fact obliged to call the sun 'most pure and most lucid,' no shadows or impurities whatever had been perceived in it; but now it shows itself to us as partly impure and spotty, why should we not call it 'spotted and not pure'? For names and attributes must be accommodated to the essence of things, and not the essence to the names, since things come first and names afterwards."

Galileo saw ruin not only in the sun but also in the moon, when he pointed his telescope there. And he much preferred a cosmos in motion and even in decay to a cosmos that, once created, never changed:

> *I cannot hear it to be attributed to natural bodies, for a great honour and perfection that they are impassible, immutable, inalterable, &c. . . . It is my opinion that the Earth is very noble and admirable, by reason of so many and so different alterations, mutations, generations, &c. which are incessantly made therein; and if without being subject to any alteration, it had been all one vast heap of sand, or a masse of Jasper . . . wherein nothing had ever grown, altered, or changed, I should have esteemed it a lump of no benefit to the World, full of idlenesse, and in a word superfluous, and as if it had never been in*

> *nature; and should make the same difference in it, as between a living and a dead creature: The like I say of the Moon, Jupiter, and all other Globes of the World.*

And this has been the drift of science ever since, in the discovery of the deep geological layers of the earth, and the vast numbers of species that have gone extinct, to be preserved only within those layers; and in the lives and deaths of the stars; and in the cycle of the life and death of the universe itself—all of which those sunspots prefigured.

If anything, the cosmos of science is as ephemeral as the cosmos of Buddha, who founded a religion on evanescence, as on a rock. Siddhārtha Gautama, who became the Buddha, wearying when very young of the sights and dread of mortality, shocked by his first sight of an old man by the side of the road, left Lumbini on a pilgrimage into the mountains:

"Grieve not for me," he said, "but mourn for those who stay behind, bound by longings to which the fruit is sorrow . . . for what confidence have we in life when death is ever at hand? . . . Even were I to return to my kindred by reason of affection, yet we should be divided in the end by death. The meeting and parting of living things is as when clouds having come together drift apart again, or as when the leaves are parted from the trees. There is nothing we may call our own in a union that is but a dream."

Mortality is the central fact of our lives. Contrary to rumor, we do know it even when we are young. We are adept at pushing the thought away, but it is there with us almost from the beginning. There are times in every life when we find it hard to think about and impossible to withdraw from. We try to number our days, so that we may apply our hearts unto wisdom, as we are advised in the Psalms. And it is essential to us at any age to know or to guess roughly where we are in our time—because that knowledge does teach us how to live. Laura L. Carstensen, a psychologist at Stanford, has presented an interesting paper in *Science*, "The Influence of a Sense of Time on Human Development." When we have reason to believe that we have decades ahead of us—our whole life ahead of us, as we say—we focus our energies on adventures, new experiences, learning new things: the advancement of learning. When we believe that we have very little time left, we focus more on experiences that have emotional meaning for us; the meaning we have found and made.

For Carstensen and her colleagues, this helps make sense of what psychologists have sometimes called the "paradox of aging." Older people tend to want to spend their time within a small social circle of a few close friends and loved ones. They want to focus their time and energy where they have already found their greatest satisfaction. And though their world is smaller, they often say they are as happy as young people, if not happier.

Ask old people how they want to spend time, and almost always that is what they say: they want to spend it with their loved ones. Young people asked the same questions will choose to spend time on new experiences. In one test, Carstensen showed people a travel poster with the usual spread of photographs: a cheetah, a parrot, a family picnicking on a trip, the Sphinx. One poster carried the message, "Capture those special moments." The other poster read, "Capture the unexplored world." The old people in the study chose to capture those special moments; the young people were more attracted to the unexplored world. "Young or old, when people perceive time as finite," Carstensen writes, "they attach greater importance to finding emotional meaning and satisfaction from life and invest fewer resources into gathering information and expanding horizons." And when we see time as virtually infinite, our priorities reverse.

In one experiment, Carstensen and her colleagues asked their subjects to imagine that their doctor had just called to say that science had made a medical breakthrough, which would give them many more years to live. Now they were willing and eager to spend time with new people and broaden their horizons. But if they were asked to imagine that they would soon leave their homes and move somewhere very far away, most of them said that they would spend their remaining time with a few of the people they were closest to. Young and old had the same reaction. What mattered here was not how old they were, how much time they'd lived, but how much time they thought they had ahead. Carstensen writes, "Preferences long thought to reflect intractable effects of biological or psychological aging appear fluid and malleable."

When we're in the first few ages of man, the last few ages seem very far away. But we do know those last ages are there, and death is there. And it is healthy and adaptive to know that; to number our days, that we may try to be wise; even if we are adept a moment later at pushing the thought away.

When we arrive at the late ages we are still consumed with the problem of mortality and still adept at pushing it away. In fact, when we're old (having arrived at that state as if suddenly), mortality means so much to us that it might crowd out everything else, if we weren't so good at thinking about it and then trying to ignore it again. An old *New Yorker* cartoon shows a man of a certain age reading the obituaries and thinking: *Twelve years older than me. . . . Five years older than me. . . . My God, exactly my age. . . .* People have computed that way since the days of the first newspapers, sometimes with a frisson of fear, but often with a strange feeling of comfort afterward. As Dr. Johnson observes, "The computer refers none of his calculations to his own tenure, but persists, in contempt of probability, to foretell old age to himself, and believes that he is marked out to reach the utmost verge of human existence, and see thousands and ten thousands fall into the grave."

From beginning to end it is this knowledge of the limit, the endpoint, death, that looms largest in our calculations and our struggles, and touches us most deeply in the stories of the struggles of our heroes. At the Eagle, patrons often wander under the blood-red ceiling of the RAF Room and read the initials of the pilots, the nicknames of their squadrons and commanders. "Donald Jimmie Moore." "Bert's Boys." "The Pressure Boys." You can make out the form of a woman who floats across the ceiling like a constellation. She is remembered in the pub as Ethel. She may have been the landlady's sister. Apparently the young airmen lifted her up to the ceiling one night and drew her outline in lipstick, and apparently she had lost her clothes.

The young airmen wrote up there with their lighters, with that lipstick, with candles, and with charcoal from the fireplace.

"Alis Nocturnes," a motto: "On the Wings of the Night."

"58." The Fifty-eighth squadron was commanded by Sir Arthur Travers Harris, known as "Bomber" Harris to the press and as

"Butcher" Harris to his men. The men were as young as seventeen, but they knew.

In what is now the Eagle's DNA Room, James Watson was oppressed by a sense of his own mortality. Watson was convinced that great scientists achieve breakthroughs by the age of twenty-five, and he barely made it. Soon after his eureka moment with Crick, and their victory lunch, Watson made his way to Paris, where he did not have much luck finding girls, in spite of his bohemian long hair and sneakers. He ends *The Double Helix* on a melancholy note, staring at the girls near Saint-Germain-des-Prés: "I was twenty-five and too old to be unusual."

Among the innumerable things it does to us, mortality binds us in mutual piety, when we're young, like those pilots. The problem of mortality pushes us to choose a path; it goads us to accomplish something, like Watson and Crick. We push it away with all of the missions that fill the first ages of man, but we know the problem is there, and it goads us to ask the largest questions—questions of ultimate meaning; questions we might never think to ask if we had all the time in the world.

This problem of mortality will define our next years and decades. It will involve not only writers, philosophers, and biologists but also sociologists, economists, and politicians. It will weigh on our minds at least as much as at any time in history, with or without the discovery of an elixir of youth. Because of our success on the planet, we face a new era even without the elixir.

"Very long lives are not the distant privilege of remote future generations," according to an analysis by the Danish gerontologist Kaare Christensen and colleagues; "very long lives are the probable destiny of most people alive now in developed countries." Life expectancy has been rising on a straight line for more than 165 years. This linear progress "does not suggest a looming limit to human life span," they argue. "If life expectancy were approaching a limit, some deceleration of progress would probably occur. Continued progress in the longest-living populations suggests that we are not close to a limit, and further rise in life expectancy seems likely."

Life expectancy has doubled over the past two hundred years, and in the last half century most of that rise came from improvements in the lives of the old, whereas before, it had come from improvements for the young. The number of centenarians on the planet has more or less doubled with every decade since 1960. At the moment, Japan is the country that offers the most years of life to its citizens. In 1950 a woman in Japan who had just reached age 65 could expect to live another thirteen years. Fifty years later, a Japanese woman who reached age 65 could expect another twenty-two years. In 1950, her chance of reaching age 100 was less than one in a thousand. By 2002, her chance was one in twenty.

According to Christensen, the elderly in Denmark are living longer without spending more years sick, frail, and in pain. A recent study followed more than two thousand elderly Danes. Between ages 92 and 100, the number of those who could live independently, shopping, cooking, and bathing, declined only slightly, from 39 percent to 33 percent. Even at age 100, one in three Danes was still independent. That's pleasant news for warm, fallible human computers of a certain age (although it's not quite as comforting as it sounds, because most of those 92-year-olds never made it to 100).

These forecasts could be wrong. The long rise in life expectancy through history has been broken here and there, chipped into jagged and serrated edges like a flint knife, interrupted by the great wars, famines, and epidemics. In the fourteenth century the Great Plague killed nearly half the population of Europe. Baby boomers have to look only one generation back to remember the global cataclysm that brought them into the world. More than 50 million people died in World War Two. Russian men died in such numbers during the war that there was a shortage of able-bodied males in the Soviet Union for a whole generation. Lately the life expectancy of Russian men has been declining again, because of too little work, too little food and medicine, and too much vodka and tobacco.

S. Jay Olshansky, a well-respected demographer at the School of Public Health at the University of Illinois in Chicago, doubts that we can ever extend the average human life expectancy beyond about eighty-five years. Even in the world's most advanced countries, with the most sophisticated twenty-first-century medicine, average life expectancy at age fifty will not exceed thirty-five years, he writes, "unless major breakthroughs occur in controlling the fundamental rate of aging." What is worse, Olshansky warns that in the United States, life expectancy may soon begin to decline, as in Russia, because of too many burgers and fries.

But more optimistic demographers, including Christensen, believe that our life expectancies will rise throughout the twenty-first century, onward and upward.

The implications of these changes for the world's economies are very mixed. In Italy and Spain today there are now almost twice as many old people as young people. With so many old and so few young, those countries and many others may be in for hard times during the next few decades. As one population expert in Washington has put it, you can't keep going with the pyramid of civilization standing on its vertex. You can't run a village, much less a country, if most of your people are in nursing homes. Chekhov wrote a short story about a coffin-maker that begins, "The town was small, worse than a village, and in it lived almost none but old people, who died so rarely it was even annoying. And in the hospital and jail there was very little demand for coffins. In short, business was bad."

If we are going to turn our population pyramid upside down in the next decades—and that's what will happen, if it stands at all—then we are looking at a highly unstable situation, socially and politically. "A civilization has the same fragility as a life," said Paul Valéry. Challenges to civil values, if they are too great, can lead to civil wars. What happens as the baby boomers go gray all over the world and have to be carried on the backs of their small number of adult children? The lengthening of our life span is the crowning achievement of our species, but the crown is heavy and the head that wears the crown is gray.

Global graying will be one of the great challenges of this century. Demographers will argue about the details for the rest of their lives and ours, just as climate scientists will argue about the details of global warming. But about the very broadest features, there are very few skeptics. We are living longer and staying healthy and vigorous later in life, and every man and woman on the street knows it. You can monitor global graying in your own hair. You can time it by the watch on your own wrist. Barring an apocalypse, the generations of humans alive today can expect to live longer (at least a little longer) than any generation before us.

No matter what else happens with the science of aging, more and more of us will follow it as global graying advances.

Since the problem of mortality will be so much on our minds, whatever our age, we will be watching this science from all sides. We'll argue not only the feasibility of its goals but their desirability.

In France during the summer of 1783, Benjamin Franklin watched the brothers Montgolfier go aloft in a hot-air balloon. "It diminish'd in Apparent Magnitude as it rose," he reported afterward in a letter to the Royal Society, "till it enter'd the Clouds, when it seemed to me scarce bigger than an Orange, and soon after became invisible, the Clouds concealing it." A man in the crowd asked, "What's the use of that?" And Franklin replied, "What is the use of a newborn baby?" He understood that the rise of modern science would mean life itself to us, and although he could not know how far or fast it would travel, he did foresee that the global enterprise would carry us toward ever longer and healthier lives. Later that same year, when the Montgolfiers staged a balloon demonstration before the French court and about 130,000 other onlookers, one Madame d'Houdetot had the same prophetic thought, and found it poignant. Madame d'Houdetot reflected, as she watched the balloon passing over Versailles, "Soon they shall discover how to live forever . . . and we shall be dead."

Now, a generation or two after that *New Yorker* cartoon with the anxious reader of obituaries, we have Web comic strips like XKCD ("A Webcomic of Romance, Sarcasm, Math, and Language"). One strip shows a line of stick figures marching up a hill toward "The Uncomfortable Truths Well." The figures suggest an endless, eternal line of pilgrims bearing questions, which the well answers one by one. And the first question? We know what it is, of course. Quoth the well: "Science may discover immortality, but it won't happen in the next eighty years."

Do we want the science to move faster? Do we want a cure for aging? The question of desirability is going to be hard for us. When we examine it closely our thoughts get tangled in it, much as we are entangled with mortality in our bodies. The spiritual and emotional knots are as tight as the biological. We're mortals. We've wrestled with the problem of mortality for thousands of years in the darkest passages of Scripture and philosophy. Our poets and artists move us profoundly by struggles without answers. No other scientific program raises so many enormous and imponderable questions, and they are so blithely dismissed by the engineers who would build the dam in the valley of the shadow of death.

We can't know yet if a cure for aging is almost within reach, if it is now low-hanging fruit. But when we turn from feasibility to desirability—when we let ourselves think about science and immortality in the same sentence, and take it seriously, even for a moment—we run into extraordinary turbulence as soon as our thoughts are aloft. Powerful currents run in us, alternating currents of yes and no. We meet internal resistances just as strong as in the body or the cell; and we only half understand them, even though we have been exploring the question "Should we?" for as long as the question "Could we?"

In *Paradise Lost*, Milton reminds us that we failed to make ourselves immortal when we reached for the low-hanging apple; in fact, we made things infinitely worse.

> *Of Man's First Disobedience, and the Fruit*
> *Of that Forbidden Tree, whose mortal taste*
> *Brought Death into the World, and all our woe . . .*

Milton reinforces the point of the lesson by making Satan fall and suffer at least as horribly as Adam and Eve. He does show some sneaking sympathy for the fallen angel, as Blake observes in a famous line in "The Marriage of Heaven and Hell"—"NOTE: The reason Milton wrote in fetters when he wrote of Angels & God, and at liberty when of Devils & Hell, is because he was a true Poet and of the Devil's party without knowing it." Even so, as Milton announces at the close of his grave and august first verse, the purpose of this epic, the point of "this great Argument," is to "assert Eternal Providence, and justify the ways of God to men."

All of Hebrew and Christian Scripture makes this assertion and insists that the ways of God are just; to be accepted with or without understanding; to be accepted even in the face of horror. Think of the ghastliest story in the book of Genesis, the testing of Abraham: in Jewish tradition, the episode of the Torah that is recited on the first day of the new year, again and again. Abraham and Sara have a child in their old age, a child so long prayed for and despaired of that when at last he is born they name him Isaac, which means "He laughs." And God comes to Abraham and commands him to take his son, "your only-one, whom you love, Isaac," up to the mountain and sacrifice him. So early in the morning Abraham saddles his donkey, takes Isaac, splits wood for the sacrifice, and with them goes up to the mountain. They climb the mountain, Isaac carrying the wood and Abraham the torch and the knife.

"Here's the fire and the wood," says Isaac, "but where is the sacrifice?"

And Abraham answers, "God will provide."

When they come to the place, Abraham builds the pyre, binds Isaac on top of it, and stretches out the knife to slay his son—but God stops him. "Abraham lifted up his eyes and saw a ram caught in the thicket by its horns. And he offered up the ram in place of his son."

Homer tells the same story of another patriarch, King Agamemnon. When Agamemnon wants to sail for Troy, the wind will not come up to fill the sails of the ships. A priest tells him that it is the will of the gods that he sacrifice his firstborn daughter, Iphigeneia. So Agamemnon dispatches a messenger to the girl's mother, Clytemnestra, and tells her to send their daughter to him. He says Iphigeneia is to be married to his greatest warrior, Achilles. And the girl comes. In Homer, the king sacrifices his daughter; but in Euripides's play *Iphigeneia at Aulis*, a goddess spirits the girl away at the very last moment and substitutes a deer, which Agamemnon kills instead.

It's curious that this same tortured story should reappear at the core of several religions. In Christian tradition, the hill where Abraham bound Isaac and lifted the knife was Golgotha, also known as

Calvary. That is the hill up which God sent his own son, carrying the wood of the cross on his back, to be sacrificed for the sake of all of humanity, as symbolized by the lamb of God.

Hindus know the story from the Upanishads. A father, Vajasravasa, pledges to sacrifice all that he has in return for the blessings of heaven. His young son Nachiketas watches as Vajasravasa's cows are led away. "Dear father," the son asks, "to whom wilt thou give me?"

His father is silent.

"Dear father," he asks again, "to whom wilt thou give me?"

Silence.

"Dear father, to whom wilt thou give me?"

"I shall give thee unto Death!"

So the boy descends to the realms of Yama, who is Death. There he learns all the paradoxes of mortality and immortality, in some of the most celebrated poetry of Hindu scripture, which concludes, "When all the ties of the heart are severed here on earth, then the mortal becomes immortal—here ends the teaching."

Framed this way, as the sacrifice of the child by the father, the story is even harder to accept than the sacrifice of the self. For anyone who has a father, or a child, it is infinitely more painful and bewildering to contemplate than the failure of Gilgamesh, or the fall of Adam and Eve. The very horror of the story forces us to reflect on the ultimate reason for this sacrifice. Whatever else they do, these stories push us to explore in the strongest possible form the struggle within us between acceptance and defiance, defiance and acceptance, in the flow of the generations. Every father does pass the problem of mortality to his child, because he must. Every child receives the problem of mortality from the father, because he must.

And framed as it is, this recurring story makes life's demands for acceptance and resistance impossible to decide for ourselves, impossible to resolve through reason, too much for mortal minds. This is the way it must be, the story says; we have to take it on faith.

The same story reappears in at least one more tradition: Lucretius retells the story of Agamemnon and Iphigeneia at the beginning of his epic poem *The Way Things Are*, a heroic effort to replace the epics of Homer and religious faith with the epic of what we now call science. In Book One, at line 101, we find the battle cry of the rationalists: *Tantum religio potuit suadere malorum*, "See what evils are done in the name of religion." Lucretius's furious line became a slogan of the Enlightenment and made him a favorite poet of the birth of science. The Loeb edition notes: "Voltaire, an ardent admirer of Lucretius, believed that line 101 would last as long as the world."

Lucretius thought he could reason his way out of these terrors. Voltaire thought science could get us past them too. The modern age of the Enlightenment would dispel all darkness. But when we approach these questions now, through secular science, they are deep as ever. The problem of mortality does not go away when we look at it from a scientific point of view. The sacrifices are real and have always been real, our inheritances of loss, borne by each generation; and now we approach them from a new direction.

Some demographers predict, for instance, that we would want fewer children if we lived for hundreds or thousands of years. We see the trend already in the world's developed countries; the longer we live, the smaller the families we choose. The trend might increase with our life expectancy. Those alive would stay alive. Those unborn would stay unborn. Galileo observed something like this centuries ago when he mocked the folly of people who think they can buy eternity in gems. He had nothing but contempt for the romantic idea that rubies and emeralds are pieces of immortality, that "diamonds are forever." All of these dreams are ways of escaping for a moment from our mortal bodies, for getting off—in our imagination—from a mortal planet circling a mortal star. Fools, said Galileo, "are reduced to talking this way, I believe, by their great desire to go on living, and by the terror they have of death. They do not reflect that if men were immortal, they themselves would never have come into the world."

Mortality is sacrifice. And the great argument of Scripture and *Paradise Lost* has its parallel now in a busy field of research, the study of the origins of aging at the level of single cells. A cell that reproduces by splitting in two will do better in the end if it divides unequally, with one half getting all the new parts, and the other half keeping some of the old parts. Cells began doing this very soon after the origin of life itself, more than three billion years ago.

According to present thinking, it all began with that first sacrifice. That was the moment when life invented aging. Those were the first cells doomed to age and die. From that moment on, mortality was ingrained within us.

Mortality is doubly ingrained in us, because it arose not once but twice. It was discovered first by those single cells, the authors of the first sacrifice, all those millions of years ago. And the invention of aging was so successful that life remained single-celled for two billion years—that is, for two-thirds of the time that there has been life on Earth. Even today, most of the life on the planet is still in the form of single cells.

Then, a billion years ago, for reasons that nobody understands, some of the single cells began to come together to form multicellular bodies. Some of the first colonies that formed were the ancestors of today's sponges, which are very simple colonies. They are essentially immortal. Their aging is negligible. Other early colonies were the ancestors of today's cnidarians, another large branch of the tree of life, which includes the hydra. The hydra lives in freshwater, but most cnidarians live in the sea, including sea anemones, corals, sea nettles, sea pens, and sea wasps—which are the world's most poisonous animals; their sting can kill in less than three minutes. The cnidarians include jellyfish and the Portuguese man-of-war. They have nerves and muscles, and some of them have eyes. But most of these thousands of species hardly age. Like the sponges, they can regenerate from a tiny piece, sometimes even from a few scattered cells. When sponges and cnidarians grow new cells, they just slough off the old ones. Any wastes that have built up in those old cells are gone, and the new cells start afresh. So the cells in those animals age and die, but their bodies live on and on.

These immortalists evolved early; they were some of the first multicellular animals on the planet. And then mortal animals evolved. Why? What advantage did they get from becoming mortal?

A sponge has no nervous system. A hydra has networks of nerves but no brain. Both animals shed their nerves the way they shed the cells that make up their skin and their muscles, and then grow new ones. The forests of delicate synapses, which tie all our long-lived nerves together in the bundles we call the nervous system, were among the most important inventions in the history of life. They allowed animals to store more and more information. Long-lived neurons allowed them to maintain a historical memory, to learn from their experience and carry experience forward. The hydra loses its memories along with its old cells. Its memories go up in the flame and ash of the Phoenix. That is a price it pays for the gift of being born anew. Although the hydra lives much longer than its nerves, it sheds its experiences with them—whereas the nerves in a nervous system can last a lifetime, and with them, we have the memories of a lifetime.

Nerves are cells, and all cells accumulate wastes and damage. They age, but they are so specialized that they can no longer be replaced. The cells in the bone marrow proliferate as long as we live, as do the cells that line our guts. They divide and divide, and any junk that's built up in them is diluted again and again so that they stay clean. In this way, the bone marrow and the linings of our guts and the cells of our liver can be said to be virtually immortal, like the hydra. But the highly specialized cells of our brains are mortal, and so are the cells of our hearts.

In essence, then, that was the second beginning of old age and mortality, in the evolution of these specialists. Ever since, animals with those kinds of long-lived but mortal cells have accumulated damage, and eventually they have failed. Because key parts of our bodies cannot last, we do not build the rest of our bodies to last. Ultimately, then, the cells that give us our identities are the ones that bring us down to the grave.

Terman and Brunk, the authors of the Garbage Catastrophe hypothesis, are among the gerontologists who have advanced this argument: that the need for the nerves brought death into the world a second time. They argue that our long-lived muscles may also have played a part in this second invention of mortality. What we call muscle memory emerges from a combination of the complex patterns we have laid down with our muscles and the firing of our nerves. It may be that the spectacularly complicated and graceful behavior of the more complex animals owes a great deal to their long-lived muscle fibers and long-lived nerves, lasting as long as the body itself.

This invention may have allowed the amazing diversification of life-forms that we call the Cambrian Explosion. If so, the invention of aging, the feature that ingrains our mortality in our flesh, made us such a success on the planet Earth.

The development of those long-lived cells would also have precluded reproduction by budding, which is the main way that hydra makes another hydra. It would have driven the evolution of the separation of bodies into the disposable soma and the protected germ cells, the sex cells. And so it would have furthered and spurred the evolution of aging. And it would have made possible something else, too. Animals with dangerous lives would have grown up fast and reproduced fast before they died. About half the animals on this planet are short-lived insects. But animals that found their way into protected niches could afford to slow down. Then they could benefit more and more from their long-lived muscles and memories. They could grow more and more intelligent. One animal line that did this more than any other was our own, the species *Homo sapiens*. We lost the gift of living more or less indefinitely, of aging negligibly. We lost the gift of living more or less negligently, without being aware of our losses. But we gained the gift of memory, of memories that can last all our lives.

We have what the hydra does not have. We have a sense of ourselves that goes back to our beginnings and looks ahead toward the infinitely various possibilities that surround our end. We exist, and we know we exist. But the price we pay is that we age, and that we know we age. The price we pay is that we know we are mortals.

And we must wrestle with these questions of acceptance or defiance.

Chapter 11

THE TROUBLE WITH IMMORTALITY

When we consider the problem of aging, and imagine that we might be able to cure it, that alternating current we feel consists of longings and dread. We are afraid of what we wish for; and most of our fears, like our hopes, have always cycled in us.

In Jewish legend, the Phoenix lived in the city of Luz. God had spared the city after Adam and Eve fell from grace because he wanted just one place on Earth to be safe from the Angel of Death. Neither Death nor Nebuchadnezzar with all his armies could storm the walls of Luz. Its citizens lived without war, flood, famine, fire, or fear. Their histories were miraculously complete; nothing was ever lost—not one hair, not a single name. There was no gate in the city walls; otherwise, everyone on Earth would have come pouring through. The only entrance was a secret passage through the hollow trunk of an almond tree that grew outside the wall. (*Luz* is Hebrew for "almond.")

Luz was the last secret patch of Paradise, almost a Heaven on Earth. King David, who had sung, and played on his harp, so many mournful psalms about mortality; and who had slept with a young girl in his old age to try to rejuvenate his cold, failing body—King David lived on in the city of Luz, singing psalms to Heaven, presumably with no more lamentations. All the great ones of the past were still there, living on and on, forever as they had been. Every man and woman of Luz was like the Luz bone, the coccyx—according to legend, the very last bone to decay in the grave.

And yet, every now and then one of the immortal men or women of Luz would suddenly say goodbye, escape under the wall, go out through the hollow trunk of the almond tree, and wander out into the world. The wise men of Luz wondered why these people left. The wise women of Luz talked about it into the night. They concluded that some citizens of Luz must grow tired of living—bored with immortality. Why else could they possibly want to leave the city of the immortals?

Alas, after the bored and the restless passed through the cave below the city and crawled out through the hollow trunk and walked away, no one ever saw them again. They were met by the Angel of Death, who buried them in the fields.

"People are always worrying about boredom," Aubrey de Grey told me once, "and it's a complete joke. I could perfectly well live till I was a million years old and I would never get bored of punting." And that may well be true. But it's also true that dreams of immortality have led to terrible nightmares of boredom ever since people began writing down their thoughts.

"Consider how long you have done the same thing," says Seneca; "a man may wish to die not because he is brave or miserable, but because he is discriminating."

Francis Bacon repeats the point in his essay "Of Death": "A man would die, though he were neither valiant nor miserable, only upon a weariness to do the same thing so oft over and over."

For one of Darwin's early supporters, Ernst Haeckel, the mere thought of such heavy boredom was more than enough to outweigh his immortal longings. In *Riddle of the Universe*, published at the turn of the twentieth century, Haeckel writes, "Any impartial scholar who is acquainted with geological calculations of time, and has reflected on the long series of millions of years the organic history of the earth has occupied, must admit that the crude notion of an eternal life is not a *comfort*, but a fearful *menace*, to the best of men. Only want of clear judgment and consecutive thought can dispute it. . . . Even the closest family ties would involve many a difficulty. There are plenty of men who would gladly sacrifice all the glories of Paradise if it meant the eternal companionship of their 'better half' and their mother-in-law."

There may be plenty of women who would make the same sacrifice.

The Czech writer Karel Čapek wrote a play on this theme, *The Makropulos Affair*, first performed during the winter of 1922 at the Vinohrady Theatre in Prague. The heroine is Elena Makropulos, an opera singer, 342 years old, who has aged through boredom into "frozen, soulless emptiness." Čapek, who wrote what we would now call science fiction (in his play *R.U.R.*, he coined the term "robot"), defended his tragic portrait of Elena in a note to his audience. "Does the optimist believe that it is bad to live sixty years but good to live three hundred? I merely think that when I proclaim a life of the ordinary span of sixty years as good enough in this world, I am not guilty of criminal pessimism." (Čapek died at forty-eight.) The Czech composer Leoš Janáček turned the play into an opera about the boredom of eternal life, and the philosopher Bernard Williams made it the basis of a celebrated essay, "The Makropulos Case: Reflections on the Tedium of Immortality."

"Who the hell wants to live forever? Most of us, apparently; but it's idiotic," Truman Capote writes in his essay "Self-Portrait." "After all, there *is* such a thing as life-saturation: the point when everything is pure effort and total repetition."

You can have a horror of death *and* a dread of eternal repetition. Woody Allen seems to suffer from both. He once said, "As long as they are mortals, human beings won't be totally relaxed." And he said, "I don't want to achieve immortality through my work. I want to achieve it through not dying." But he also said, "Eternity is a long time, especially towards the end."

In fact, for some people, even the life span we have now is boring; they already feel they are too long for this world. To fill the time they act like the gods on Olympus, manufacturing excitement; or like the restless souls in Luz, testing the edges of mortality. Olympus was for the ancient Greeks, as Luz was for ancient Jews, an eternal reward for some of the greatest heroes, including Hercules, who struggled all his life with the problem of mortality. Hercules rescued Prometheus from his chains on the cliff. Hercules fought with Geras, who represented hideous Old Age. Hercules wrestled with Death himself, to rescue the wife of Admetus. When the poison of the Hydra killed him in the end, he was allowed to ascend to Mount Olympus. There Hercules married Hebe, the goddess of youth. But what did the immortals do all day every day on the eternal mountain? They squabbled like mortals; they relieved their boredom by watching the mortals down on the plain. Not even the Greeks could imagine a way to escape the tedium of immortality.

Tortured mortals tie their brief time on Earth into knots. The seven deadly sins take them to the edge of the city of life, or out of it. All kinds of craziness lift us out of time, and then return us home, feeling almost reborn (if not necessarily refreshed). Even people who lead outwardly calm lives find paths to the edge. The most petty, trivial idiocies can yield that strange thrill of having flirted with the Angel of Death, just beyond the Luz tree. Procrastination is not one of the seven deadly sins, but those who work hard at it do sometimes achieve a near-death experience.

Of course, as an argument against immortality, the problem of boredom cuts both ways. Most of us have learned to deal with it already. And we all know people for whom boredom is not an issue in life—or at least, people who finesse the issue with brio again and again. A while back, a guest came to dinner who had made a brilliant success of seven careers, beginning with law, journalism, and politics. He'd served as a senior vice president at a bank and as the city manager of one of the biggest and hardest-to-manage cities in the United States. It was no Luz, and it wasn't boring. Now he spent more and more time with his grandchildren, although he'd begun yet another career on the side, as the founder and president of a humanitarian nonprofit organization. Sitting on my terrace above Broadway, he picked up a book I was reading at the time, *The*

Denial of Death, by Ernest Becker; opened it at random; and read a passage aloud:

> *A person spends years coming into his own, developing his talent, his unique gifts, perfecting his discriminations about the world. . . . And then the real tragedy . . . that it takes sixty years of incredible suffering and effort to make such an individual, and then he is good only for dying. . . . He has to go the way of the grasshopper, even though it takes longer.*

My guest laughed. He was sixty.

I said, "I'm glad you're laughing."

People like that might find good things to do if they had a thousand years. On reflection most of us can think of at least a few things we'd like to do. A playwright wrote to me after we talked about it. At first he'd been horrified by the thought of boredom. Now he allowed, "Time to read everything would be one of the consolations of immortality."

But I think boredom may be merely an intimation of much deeper fears about our oldest dream. Most immortalists assume that we would attain a state of maturity, presumably of young maturity, and stop there for centuries. In other words, if we engineer a state of negligible senescence, we will cease to travel through the seven ages of man. We might choose to stop at the third age, the age of the lover, "Sighing like furnace, with a woeful ballad / Made to his mistress' eyebrow." Or we might linger at the fourth age, the age of the soldier, "Full of strange oaths, sudden and quick in quarrel, / Seeking the bubble reputation / Even in the cannon's mouth." Elena Makropulos, the opera singer, spent 300 years at the age of forty-two.

Wherever we stopped, the fact of having stopped there would do strange things to us, because none of life's stages makes much sense in itself; each makes sense only in relation to the next and the next, only in a series. Without the sense of a progress through life, life is a kind of stasis. Identity itself is stasis, or it would not be identity. My guest on the terrace had come as close as anyone I know to having a series of lives, hopping from one to the next—but he'd kept his family and friends, and he'd kept himself. That is, he'd been recognizably himself in every one of his careers. Our character is fixed because we do so much imprinting in our first, second, and third ages. In our first age, we imprint on language. In our second age, we imprint on music. In our third age, we imprint on the work of a lifetime, if we're lucky; and on the love of a lifetime, if we're very lucky. And after that? Then, the working out of the plans. It's all arranged, in some sense. Life after those first three ages feels like an unfolding, a development, although we do have to build the particulars of each new stage with infinite labors and pains. As we grow and live and choose, always with a sense of discovery, we build on the givens; we work from our first premises, pursuing those first loves. The rest of the seven ages are, in effect, a playing out of an identity that has become, in most ways, permanent and indelible.

So how would it work out if we had a thousand years at one of those seven stages? In life as we know it, each stage is a waypoint on the way to the endpoint. A huge part of the action and the drama in the seven ages comes from the sense of an ending, the knowledge that all these ages must have an end. "Immortality, or a state without death, would be meaningless," argues Williams, the philosopher, in his essay "The Makropulos Case," because "death gives meaning to life."

Again, we're talking here only about the deepest problems of existence. They make even the philosophers shrug, or crack jokes. In his essay, Williams quotes Sophocles: "Never to have been born counts highest of all"—and the wry old Jewish reply, "How many are so lucky? Not one in ten thousand."

Nevertheless, it is true that most of us wend our way through the seven ages with a sense that we are making a journey in the world. Each of us feels like one pilgrim on one journey, lost or found. That is why Dante could capture each of his characters in a single image, in the *Inferno*, the *Purgatorio*, and the *Paradiso*. He saw the people he had known in his native Florence as they were in eternity, quintessentially themselves. (The critic Erich Auerbach calls Dante the great poet of the secular world.) If each of those Florentines had lived in the flesh for an eternity, or even for a mere thousand years, could any of them still say they had been true to their first loves? Would their lives have retained any shape at all? Elena Makropulos's problem wasn't being forty-two, Williams writes. "Her problem lay in having been at it for too long." She was bored because "everything that could happen and make sense to one particular human being of 42 had already happened to her." At least, everything that could happen to and make sense to a woman of a certain character had happened. Her character had long since formed; her destiny had been formed by her character. Nothing remained but to live it out over and over again, like an endless performance of exactly the same song. We are performers of the self, we are playwrights of our lives, and we need death to bring down the curtain, or the play will go on too long; the story will lose all shape and cease to be a story at all.

In short, we are afraid that we will gain a world of time and lose our souls.

During my summer in London I looked up Martin Raff, a distinguished biologist at University College. Raff was an energetic, charismatic man whose talk was powerful medicine for the metaphysical hangover that I got whenever I talked with Aubrey. Raff leaned farther in the opposite direction than anyone I'd ever met.

Raff trained as a doctor and then switched to biology. After a celebrated research career in cell biology—during which he did valuable work in immunology; explored the structure of cell membranes, the growth of young neurons, and the lives of stem cells; and won many prizes and honorary degrees—Raff retired in 2002, a little before his sixty-fifth birthday, because he felt that when it was time to step down one should make way for the next generation. As he said in his retirement speech, he could never understand why people would want to live beyond their allotted time. He had worked at the National Institute for Medical Research, in London, when the institute was directed by Sir Peter Medawar, the originator of the evolutionary theory of aging. Raff didn't want to be like Medawar, who had fought against retirement to the end.

Raff himself had been happy to retire, he told me, although

I knew that he'd enjoyed a charmed career as a scientist. Among other things, he was known for his studies of a cellular process called apoptosis. Our cells are constantly receiving signals from cells around them and from their own innards. Somehow these signals tell them when it is time to go. When that time comes, they die. A cell that fails to die can start a cancer, but almost all cells do heed the signals. Apoptosis is cell suicide.

One morning, we had tea. Raff is tall, lanky, gray, lined. He wore eyeglasses with wireless rims, sandals, jeans, and an old, comfortably tired, faded button-down shirt. He had never heard of Aubrey de Grey. When I showed him a copy of Aubrey's journal, *Rejuvenation Research*, he turned the pages with a small, fond, indulgent smile as if to say, "We were all young once." He cackled softly here and there as he skimmed Aubrey's opening editorial. He read aloud: *"Aging has been with us for a long time. The idea that it will be with us forever has ceased to be tenable."*

"Oh," said Raff.

I'd put out croissants from a pastry shop in England's Lane. When I apologized for handing him a chipped cup, Raff laughed and told me not to worry. He said, "England is the Land of Chip." I saw a chance to press the topic that we had met to talk about, and quoted W. H. Auden:

> *And the crack in the tea-cup opens*
> *A lane to the land of the dead.*

Raff simply laughed again, as if I had just handed him, as a gift, the perfect riposte. He quoted a few lines from one of his favorite songs, "Anthem," written by an old friend of his, the poet Leonard Cohen:

There is a crack—
A crack—
In everything.
That's how the light gets in.

Over tea, Raff told me that he did not see aging as a very interesting biological problem. In his view, the problem was solved. After all, he said, a hydra doesn't age. (He meant the real hydra, the cnidarian, the modest little pond creature with the waving tentacles—not the mythical Hydra, the monster that Hercules slew.) The reason that a hydra does not age is that its cells do not hang around for very long. The cells are always being newly generated and sloughed from the tips of the tentacles. If all of our cells and macromolecules were turning over, then we would not age, either, because the oxidative damage to those cells and molecules would vanish. "But unfortunately it ain't like that in most animals—or plants, for that matter," Raff said. "And I would be pretty pessimistic that you could stop oxidative damage or defend against it forever. It seems quite unlikely."

And he did not find that thought depressing. "You do have a sense of what your life is, and what you want it to be, and what at the end you'd like it to be," he said. "And whether you're able to do it is really just luck. I mean, at every level it's luck. Because, you know, you have to have the right genes, you have to have the right environment. Your kids have got to be not knocked over by cars. There are just so many imponderables that are out of your hands.

"And so at the end of your life you have to build into it the slowing down. Because no matter what you do, unless you're a hydra, you're going to slow down. And once you think of that, then you need that end to be incorporated into your life plan. I think it's helpful to have a plan. You've got to be very very lucky to be able to carry out the plan. It's all luck. But nonetheless.

"The end of life is just as important as any other stage in life. Of course, it's something most people fret about. And to spend a lifetime or even a third of your lifetime fretting about it is bad. You don't want to be fretting about anything. You want to be looking forward to everything."

In his retirement, Raff hoped to promote certain causes that were passions of his. One of the dearest to his heart was euthanasia. Raff believes it should be easier in our society to help people die with dignity. His impatience with immortalists and his passion for euthanasia are attitudes that dovetail with his science, although when I asked him about it he said that he thought his work on cell suicide and his work on euthanasia had no connection with each other.

Some years before, a close friend of his, Gavin Borden, the publisher of the textbook *Molecular Biology of the Cell*, which Raff helped to write, developed cancer. Raff became closely involved with his friend's treatment, and when Gavin was near the end, in agony, he asked Raff to help him commit suicide. When Raff's parents were very old, two happy and highly accomplished people, still in fairly good health, they decided that it was time for them to go. With their sons' blessing, they committed suicide together.

Raff told me both stories at length. Because euthanasia is illegal in Florida, where his parents lived, and in New York, where his friend Gavin lived, they were both horror stories: lawyers, doctors, obstacles of a dozen kinds; anguish, torment, involuntary confinements, a shotgun under an invalid's bed.

When he was done, there was a long silence between us. At last, Raff sighed. "So that's what I meant when I said you have to be lucky. I mean, it's a crapshoot whether you end up in their position. But surely what anybody would want is the assurance that when the time comes, and you want to end it, you'll be in a position to do it. Then people could stop worrying.

"I mean, if you ask people, most people are not afraid of death. Most people are afraid of dying—of terrible dying. That's what they're afraid of. And justifiably so. I don't know what percentage of deaths are awful, but in my time as a doctor it was high. And I'll bet it's still high.

"I don't understand why more people don't feel the way I do. I wouldn't want to extend my life for a second!" Raff said. "I wouldn't want to go backward—not for a year, let alone twenty years. But people are very different. I remember Peter Medawar—he had stroke after stroke after stroke. It was just horrible. In the end he was bedridden, he was blind. . . . And he still wanted to live."

Raff said he would feel wretched if someone told him that he could now live 500 years. "That would be one of the most acutely depressing things anyone could say to me. My life has been terrific; I've been spared most of the awfulness—not all of it, but most of it. But I see life as stages. And the goal is, in every stage, to *like* it, to be lucky enough, to be healthy enough, to get through it with pleasure, and always be looking forward to the next stage. And when you get to the next stage, it doesn't disappoint you but actually turns out to be better! And then, that would include death! Why not? Why not have a life where you're looking forward to every stage, it always is better than you think, *and even the end* is just as good as you had been hoping? Why would that not be a goal?"

As a boy, he told me, he'd loved football, basketball, ice hockey, skiing, tennis, and sailing. Now he'd made a list of a hundred-odd things he hoped to do in his retirement, and he was happy about them all, including his campaign for euthanasia, and his preparations for dying.

Raff was flabbergasted that not everyone else in the world shares this ebullient view of life. He himself had made the most of each stage. He would never cry, like one of Shakespeare's bitter kings, "I wasted time, and now doth time waste me." But he doubted that he could keep on being so lucky if there were, say, seventeen ages of man.

"Now if you have this view of it, to extend life to five hundred years—you're asking a lot, now!" Raff laughed a mortal laugh. "You know, it's just too long to carry on this kind of thing."

So we worry that the regime of the self can go on too long. That is one of the chief reasons why we resist the idea of a cure for aging—why the question of desirability is as complicated for us as the question of feasibility.

And then of course we have to consider not only the seven ages of man, but also the ages of humankind. In history, too, regimes can go on too long. This is true in science and art, where each wave of great ones makes way at last for the next. The German physicist Max Planck said, "A new scientific truth does not triumph by convincing its opponents and making them see the light, but rather because its opponents eventually die, and a new generation grows up that is familiar with it." This is often paraphrased, "Science advances funeral by funeral." Most Young Turks of science like that quotation, including Aubrey de Grey, but it is a strange one for him to trumpet, if you think about it.

Then consider what a cure for aging would mean in politics. If emperors could live forever, we might have no freedom anywhere. At the moment they are merely figures of romance in which we see our own struggles writ large—reminders of what the situation is for all of us, no matter how we try to defy it.

In *Antony and Cleopatra*, Shakespeare gave the queen of the Nile one of the greatest death scenes in the history of drama. "I have immortal longings in me," she cries in the last act, just before she finds the asp in the basket of figs and holds it like a baby to her breast.

In the East, one of the great examples of immortal longings in high places is Emperor Wu, who was to China what Julius Caesar was to the Roman Empire. Wu presided over China's greatest expansion, during the Han dynasty. Like Caesar, he was a celebrated writer. One of his prose poems, "Autumn Wind," ends, "Youth's years how few, age how sure!"

According to legend, Emperor Wu set up bronze statues of immortals at his palace, holding pans to catch the dew from the moon and make an elixir of immortality. He did rule China for more than half a century, but he grew old at the same rate as his subjects and died in 87 B.C. Centuries later, after the fall of the Han dynasty, in A.D. 233, Emperor Ming sent his court chamberlain to cart away the earlier emperor's statues, pans and all, and set them up at his own palace. It is said that as the statues were hauled toward the carts, they wept.

Some of China's greatest artists have celebrated those legends, notably Li Ho, who had the short life of a lyric poet, from the years 791 to 817. Li Ho laments the failure of our quest for immortality without feeling the least guilty about it. His poem about the weeping of the bronze immortals is all pathos, autumn winds, and withering orchids. "His pessimism," the translator and scholar of Chinese poetry A. C. Graham observes, "has none of the ambivalence which one expects in a Western artist obsessed by original sin, who is at least half on the side of the destructive element because he finds it at the bottom of his own heart."

"This is why I fear research into aging," writes David Gems, a gerontologist at University College London who is one of the most prominent researchers in his field. If biologists could have done for the dictators of the twentieth century what they can now do for roundworms and flies—double their life span—then Mao Zedong might still be alive. Mao would be in the middle of his life, as Gems says, "and might not be expected to die a natural death until 2059." Joseph Stalin would still be alive, too, and perhaps going strong. You can argue that dictators seldom die of natural causes. But giving very bad men very long lives would not be good for the world. Thousand-year Hitlers, thousand-year Reichs. Gems sometimes remembers the words of Winston Smith's torturer in George Orwell's *1984*: "If you want a picture of the future, imagine a boot stamping on a human face—forever." "This 'forever,'" says Gems, "is what biogerontological research has the potential to achieve."

The regime of the self can go on too long, and the regime of the ruler. We can even worry about the regime of the species.

Virginia Woolf watched a moth die on her windowsill one morning as she was writing at her desk. "Oh, yes," it seemed to say, "death is more powerful than we are."

We already overcrowd much of the planet. We bestride and consume it, present and future. We eat so much more than our share that the generations following us will inherit a very poor place to live.

If a cure for aging became available to the rich before the poor, which is the way the world always turns, then the unfairness of life might become absolutely unsustainable. How would our world of haves and have-nots go on spinning if the haves lived for a thousand years while the children of have-nots went right on dying hungry at the age of five? And what would happen to the rest of the living world? Would the other species on the planet, the other earthlings, have even less? Woolf pitied the moth on her windowsill. The poet Robert Burns felt for the field mouse revealed by his plow. How often would we pause to look beyond ourselves, or stop the plow, if we lost that fundamental connection with the rest of life—tenuous as it is already—and loosed the bonds of age?

We want a good long life. We also want a good life. It's hard to see how members of our species could have both for very long if more and more of us had to make do with less and less. Still, the adventure of living another five hundred years on a planet as overburdened as ours would be, if nothing else, an antidote to boredom.

Maybe, just maybe, we would tread more lightly on the Earth because we would each preserve one body, one piece of human equipment, instead of continually having to replace it. In that sense, thousand-year lives would be the ultimate in conservation. We might even grow up faster as a species if we lived long enough to pay the price for our species's sins in our own skins. But when we talk about the health of the body and the health of the planet, we deal in goods that are difficult to reconcile.

The skin-in people, the molecular biologists, explorers of the interior, worry about the body. The skin-out people, the evolutionary biologists, students of the rest of the living world, worry about the biosphere. They worry outward. And in the end, we need both. We won't be long for this world unless both stay healthy. Although it was written in a different connection, there is a beautiful passage in one of the papers of William Hamilton, the theorist of the evolution of aging, that speaks to our situation. "Perhaps the most interesting thing to come out of the realization of possible conflict within the genome is a philosophical one," Hamilton writes. "We see that we are not even in principle the consistent wholes that some schools of philosophy would have us be. Perhaps this is some comfort when we face agonizing decisions, when we cannot 'make sense' of the decisions we do make, when the bitterness of a civil war seems to be breaking out in our inmost heart."

And then we have decisions that bear on the human genome itself. In the Gospel according to Luke, in the King James version, Jesus asks, "And which of you with taking thought can add to his stature one cubit?" This line is rendered, in the New International Version: "Who of you by worrying can add a single hour to his life?" And Jesus caps his question: "Since you cannot do this very little thing, why do you worry about the rest?" A cubit for the ancients was the distance from the elbow to the tip of the middle finger, about twenty inches. Now we live in an era when we really can add a bit, if not quite a cubit, to our stature, with the help of human growth hormone (HGH). This hormone adds inches to the stature of thousands of very short children every year. In the same way, in our lifetimes, or the lifetimes of those children, we may figure out how to add years to our lives by slowing aging. Oddly enough, HGH has been taken for that purpose ever since 1990. There's no solid evidence that it works. Even so, it is sold by antiaging companies, hawked everywhere on the Web, and recommended by the controversial American Academy of Anti-Aging Medicine, although biogerontologists denounce the academy and its claims.

What happens when we have real antiaging pills that pass the tests of clinical trials? As bioethicists have begun to note, this is a problem that would make all of our bioethical debates to date look small. What are the bioethical problems that have exercised us in the last ten or twenty years? Stem cells. Cloning. Gene therapy. The privacy of genetic information. Steroids. All of these problems matter in themselves, but all of them would be subsumed in the transformations of society and human nature that would be wreaked by a significant success with the human life span. And then will come the option of changing the genome itself. We will add or subtract genes to lengthen our lives, until there is no going back, because no human beings alive (however long they may live) will ever be human in the same way again. Then there will be no escape from Luz.

The regime of the self, the regime of the ruler, and the regime of the species. If we are going to survive to enjoy a good portion of the future, our health and happiness depend on a great deal of luck with them all. We all know this, and it is part of the alternating currents of hope and dread that we feel when we listen to the engineers of longevity. It's a mad regime that tries to make itself immortal at the cost of the world around it; as mad as a regime that surrenders life and throws it away.

There may even be some hidden likeness between the skin-ins who try to conquer aging and death, and the skin-outs who are willing to let the natural world conquer them. Either the will to power or the will toward submission can be carried to a pitch that is near madness. In all the annals of the surrender to nature in the writings of naturalists, Hamilton's essay "My Intended Burial and Why" is probably the most extreme. It is beautiful, too, in its own way. "I will leave a sum in my last will for my body to be carried to Brazil and to these forests," he writes. "It will be laid out in a manner secure against the possums and the vultures just as we make our chickens secure." That is, his body should be enclosed in a coop to keep out the larger carrion-eaters. He bequeathed it, instead, to the *Coprophanaeus* beetles. "They will enter, will bury, will live on my flesh; and in the shape of their children and mine, I will escape death. No worm for me nor sordid fly, I will buzz in the dusk like a huge bumble bee. I will be many, buzz even as a swarm of motorbikes, be borne, body by flying body out into the Brazilian wilderness beneath the stars, lofted under those beautiful and un-fused elytra which we will all hold over our backs. So finally I too will shine like a violet ground beetle under a stone."

The trouble with immortality is endless. The thought of it brings us into contact with problems of time itself—with shapeless problems we have never grasped and may never put into words. Our ability to exist in time may require our being mortal, although we can't understand that any more than the fish can understand water. What we call the stream of consciousness may depend upon mortality in ways that we can hardly glimpse.

Not long before he died, I paid a call on the eminent molecular biologist Joshua Lederberg, who, toward the end of his life, had helped lead the science of gerontology. In his twenties he'd done work in genetics that won him a Nobel Prize. In his eighties, he had been invited to serve as the chief scientific adviser to the Ellison Foundation, which became, at his suggestion, one of the world's largest private supporters of gerontological research. We met in his office in Founder's Hall, at Rockefeller University, where he had once served as president.

Lederberg still had a strong, alert stare, but his steps were feeble now. He used a walker to move around the desk and greet me. His beard and his gravity gave him a famously rabbinical look; he was descended from rabbis on both sides, and his mother could trace her family tree back through a long series of rabbis and rabbinical scholars all the way to Dov Ber, the Maggid of Mezritch—the Great Maggid, who led the Hassids of Eastern Europe in the eighteenth century.

"Did the Maggid live a long life?" I asked him.

"I have no idea," he said brusquely. "Elie Wiesel might know." Lederberg said he'd never thought much of the idea of glorifying one's heritage by tracing one's distant ancestors. "So we go back. After Methuselah, then what?"

We talked about the evolution of aging, and soon got into very deep water—or else very shallow water, because one realizes, in conversations like this, how shallow all our precepts and percepts may be.

"But exactly what is it that's being conserved, when you talk about immortality?" Lederberg asked at one point. "Do you want to freeze your identity or are you willing to die a little bit to let innovation creep in?"

Some part of you dies every second, he said, as your neurons go one by one. And a certain number of neurons are also born every second. That's part of neuronal turnover. "And the whole corpus of our memories, our recollections, changes from instant to instant. If we could do it, exactly how much of that do we necessarily need to conserve?"

He gave me a long level stare. "In other words," he said, "how much immortality do you want?"

Chapter 12

THE EVERLASTING YES AND NO

Not long ago I had breakfast with Eric Roth, a Hollywood screenwriter who lives in a beach house in Malibu. He had just finished a screenplay about a character who is born old and grows younger and younger, living the seven ages of man in reverse. This screenplay was inspired by a short story, "The Curious Case of Benjamin Button," which F. Scott Fitzgerald published in *Collier's* magazine in 1922 and included in his book *Tales of the Jazz Age*. As a newborn, Benjamin looks seventy, the biblical three score years and ten, and his father calls him Methuselah. By the time he dies, Benjamin is a baby at last, as lost to the world as the very oldest old, sans everything.

Roth's house is built right on the edge of the Pacific. You walk out the back and down the stairs, and then you have to take off your shoes. After our bagels and coffee, I borrowed a pair of swimming trunks and went wading out alone into the ocean. Roth was already upstairs on the second floor tapping away at his computer keyboard, surfing the Web, fishing for his next project; but I had to go into the water.

The bottom dropped away in just a few steps. Almost instantly I was over my head.

"The test of a first-rate intelligence is the ability to hold two opposed ideas in the mind at the same time, and still retain the ability to function," Fitzgerald wrote in one of his notebooks. By that standard, mortality itself is beyond us. We still can't hold it in our heads, although we never get tired of trying. All our lives, we're astonished to find ourselves growing older.

Even as a civilization we are simpleminded. The dream of utopian science, the cure-all of cure-alls: that is one idea we hold in our heads. In Francis Bacon's fantasy of the New Atlantis, one of the first science-fiction stories, he describes a new foundation, "the noblest foundation that ever was upon the earth," a company of brilliant minds exploring and discovering the way things are. "The end of our foundation," its spokesman declares, in Bacon's fantasy, "is the knowledge of causes and the secret motion of things; and the enlarging of the bounds of human empire, to the effecting of all things possible." Bacon hoped the brilliant sun of that New Atlantis would rise in the New World, and bring on the dawn of everlasting youth. We live by the light of that hope everywhere today, from East to West—as John Updike once put it, "our craven hope that science will save us."

When the sun shines, we dream of perfect knowledge, ageless bodies, utopian economics. The smartest people, the best and brightest, fall for fads that are, on the face of it, doomed. *Eat More, Weigh Less.* The market will only go up. The rising tide will lift all boats, with the help of the invisible hand. Ever more. We *know.* We mix spiritual and anatomical and financial advice, wisdom literature and medical literature. When we read in the Upanishads: "There are a hundred and one arteries leading to the heart; one of them pierces the crown of the head. He who goes upward through it, attains immortality," most of us are clear which magisterial realm we are swimming in. And yet we buy thousands of books that confuse our innate sense of direction; and when we try to find our way upward through them, we miss trouble down here below.

Dystopian science, the nightmare of absolute disaster: that is another idea we hold in our heads, but not at the same time.

The surge of our feelings around this problem of mortality is so simple and so repetitious that our surprises of grief and delight are endlessly renewed, like incoming waves on outgoing waves. To hold two opposing thoughts in our heads is completely beyond us. But what makes us hope the world is absolutely perfectible, from the skin out and the skin in? And what makes us despair of all possibility of meaningful improvements? Why always one or the other? Why should our fate be any simpler than our physiology?

Back on land in Malibu, Roth told me how he'd adapted the story of Benjamin Button, and how Brad Pitt had signed on to play Button, and Cate Blanchett the love of his life; how the special-effects team had managed to show Pitt as a baby who is born with wrinkles but is rejuvenated with the passage of time. When I told Roth what I was working on, he said it was obvious that biology would make us immortal soon, in another twenty or thirty years. I forget how many years he said. A cure for aging was so close and so settled a question that it wasn't even interesting. We have learned so much now that immortality is inevitable. Anyone who knows anything about the way science is accelerating has to feel that way. "Don't *you*?" he asked. "Don't *you*?"

For the Victorian critic and philosopher Thomas Carlyle, the two great contraries in life were spiritual: the "everlasting yea" and the "everlasting no." His great *yes* meant faith in God and all that is right, sacred, just, transcendentally good in the world. His *no* meant the hell of evil, unbelief, all that is spiritually withered or dead. Our souls have to choose between that everlasting yes and no.

For us—for my crowd, at least—when arguments turn cosmic, the great contraries in life tend to be material. We don't argue very often about our faith in an afterlife; but we do sometimes argue about the feasibility and desirability of another fifty years right here. And again the answers tend toward yes or no.

When you talk this way, you often hear yes and no from couples. Once, over a dinner in Princeton, I asked a distinguished writer and his wife if they would like that extra fifty years. She found the idea repulsive; he thought it would be delightful, as long as he could play with his grandchildren and ride his bicycle. He looked at her with tenderness that was oddly like an apology.

Eric Roth and his wife, Debra Greenfield, also voted yes and no. But he was a screenwriter with a movie about rejuvenation, and she was a lawyer who had just taken a degree in bioethics, so that may not be surprising.

The greatest extreme I've ever encountered was in the marriage of Aubrey and Adelaide. Like Aubrey, Adelaide wore simple old hippie clothes that had survived a lot of washings. Unlike him, she'd long ago lost every ounce of professional ambition. She had given up teaching and the rat race of publish or perish. When I first met her, she had just spent a year repeating someone else's failed experiment with a fly gene called *scant*, finding and fixing the mistakes, so that her colleague could publish the work. "I'm a good geneticist, but I have no career anymore," she told me with a smile. Most Americans take good care of their teeth, but she seemed to have let hers go, and most of them had gone. The few that were left were stained tobacco-brown.

She didn't seem to miss her tenure or her teeth. She sat at her tiny desk all day under her Gothic stone winding staircase and solved the eternal problems of scientists whose projects had hit glitches. While she worked, she chain-smoked Parliaments, Pall Malls, Marlboroughs, Camels, Lucky Strikes—anything without a filter.

Aubrey loved Adelaide very much, and it was one of the sorrows of his life that she had no interest in immortality. I often saw a softening of his hard pale face when he looked at her, a sort of warming of the alabaster. "I can't make progress talking with Adelaide," he told me sadly. He felt the most bitter frustration that he couldn't change her mind. "Whereas with others it doesn't upset me. It's their choice."

Aubrey had arranged to have his head frozen if he died prematurely, to be revived when the day of everlasting youth has dawned; and at night, over dinner, he was trying to coax her to do the same.

* * *

When it comes to our health, most of us find ways to resolve the everlasting yes and no. Whether or not we expect eternal life, or any little gift of extra time, we try to take care of ourselves.

"I am inclined to think," Descartes wrote to a friend, "that I am now farther from death than I ever was in my youth." He told a visitor, an English philosopher, that although he could not promise "to render a man immortal . . . he was quite sure it was possible to lengthen out his life span to equal that of the Patriarchs." In other words, a thousand years. And yet, when Descartes visited the young Blaise Pascal in Paris, in 1647, and found Pascal sick in bed, what did Descartes prescribe? Stay in bed, get lots of rest, drink soup. In principle, Descartes wrote in a letter, he could understand the body down to the last detail. "But for all that," he confessed, "I do not yet know enough to be able to heal even a fever."

Leonard P. Guarente, of MIT, who breeds Methuselah yeast and worms in his laboratory, takes a multivitamin when he gets home: "One of the ordinary, common ones. Also vitamin D. And a low-dose aspirin." He does not take resveratrol, the substance in red wine that the Guarente lab discovered, the drug that nurtures those yeasts in the petri dishes. Web sites are hawking resveratrol as a life-extension supplement now. The ads pop up everywhere, even next to a philosopher's essay on the acceptance of death, a priest's sermon on the afterlife, a careworn caregiver's blog about local hospice care: "Longevity on Sale. Relax. Take a deep breath. We have the answers you seek." Guarente says he will think of adding resveratrol to his daily pills, if and when the stuff is shown to work for us, and safe doses are established, and you can buy them in the drugstore. "Antioxidants, etcetera—none of it works," he tells his friends. "That's the best, the simplest thing to say. So far no drugs have demonstrated that you live longer. I defy you to name one."

What we do for our health does matter to our life span. Although genes are important—hence the wise old adage, "Choose your parents well"—behavior matters more. Studies of identical twins suggest that more than two-thirds of the variability in our life span depends on our environment; and environment, if we have any say in it, if we are free to make choices, is what we ourselves make of our lives. But generally what we should do is what we have known about for centuries. Listen carefully to the immortalists and then make yourself a good cup of soup. A long review article on the latest work in stem cells concludes: "The best advice is still to eat moderately and exercise moderately." Tom Kirkwood, reviewing Raymond Kurzweil's book *Fantastic Voyage: Live Long Enough to Live Forever*, notes that most of its advice about health is sane, sensible, and very familiar. To enjoy the fantastic voyage into the future, stay with the tried and true.

The house where we lived during our summer in London, with the little garden in front and the wisteria that twined around the door as if to say, *There will always be an England*, was owned by an old friend of mine. He'd had a career as a publisher, and he was working on a novel, but he spent most of every day as a househusband. His life was a cross between our contemporary dream of fatherhood and the old dream of the English gentleman. His library was lined floor to ceiling with the world's great books, but also with the latest instructions on how to live forever. He ate well, he kept fit, he had a first-rate intelligence, but he also took dozens of antiaging pills every day, mail-ordered from Florida.

One morning he had a talk about aging with his youngest son, who was about to turn nine. He said we know enough now to live to the age of 120, at least. That day he planned to have lunch with an old friend who needed cheering up. He left the house at midmorning and jogged to his gym. A few steps from the entrance, he fell down as if he'd been struck by a fist. It was a heart attack. An ambulance came within a few minutes, but he was dead.

Should we drink wine for our health? That's unclear. Should we drink red wine rather than white wine? That's unclear. Should we buy resveratrol? If you want to live long enough till medicine knows how to save you, stick with what medicine knows how to do well now. It's healthy to remember where we are in the scheme, in the immense journey of the caravan. Aubrey ends his book *Ending Aging* by saying, "Eat well, exercise, and support the Methuselah Foundation." He will look forward to shaking your hand someday in the distant future, with "the dark specter of the age plague driven away by the sunshine of perpetual youth."

That is fairly close to the health advice that Descartes gave Pascal.

Once after Aubrey and I visited Adelaide's nook, the three of us trooped up the narrow spiral stairway to the roof. Adelaide often nips up there for a smoke among the weather vanes of Cambridge: the cock, the fish, the bronze cupola, exposed pipes, stone towers, and ugly tallish buildings that looked as if they were wrapped in aluminum foil for the freezer. She smokes by her rooftop garden, a few tiny clay pots. There we stood, the immortalist, the geneticist, and the journalist, admiring the same humble herbs that mortals have applied to the same problems for more than a thousand years:

the leaves that may have done a little good, the sap that at least did no harm. "And this is aloe vera," said Adelaide, "a salve for my poor hands."

The field of gerontology is also divided between yes and no. Will aging be with us forever or not? Should gerontologists try to cure it, or just make our final years less awful? Are students of mortality poor cousins at the fringes of medical science, or are they working toward the salvation of the world, building the fountain of youth?

Gerontologists can't even agree if aging is a single unitary problem. Lederberg told me, "I'm still struggling to decide if there is a biology that can be called aging, that's different from what we call developmental biology on the one hand, and on the other—how shall I say this—simply the accidents of existence. If nothing else, the truck will run you over. With a certain probability. You can do statistics on that, and that's part of aging as well."

In other words, there are the accidents that strike our bodies from outside, and the accidents from inside—the accidents of metabolism. "Rusting," Lederberg said simply. If that is aging—if aging is nothing but the accumulation of all those accidents—then every medical program, from the pediatric to the geriatric, is a campaign against aging (and so is traffic safety). Current disputes in gerontology may come to seem less in the realm of the metaphysical and more in the realm of the methodological. We may see a reconciliation of the radical and the conservative gerontologists. Some of the field's leaders, including Robert Butler, the first and founding director of the U.S. National Institute of Aging, who is now eighty-two years old, are talking about the goal of extending human life expectancy by seven years. One of the field's leading advocates, Dan Perry, argues that "the wellsprings of scientific knowledge capable of constructing new dimensions of life, health, and longevity are waiting to be tapped in the twenty-first century."

O Well of Gilgamesh!

Ana Maria Cuervo, the lysosome expert, collaborates with Dave Sulzer, a neuroscientist at Columbia. Cuervo and Sulzer are working on coaxing lysosomes to do better with the junk that causes neurodegenerative diseases. Sulzer says he and most medical researchers still feel it is more urgent to work on cures for specific diseases like schizophrenia, autism, Huntington's, and Parkinson's than to work on aging. That's the worldview they grew up with, he says. "Aging still seems more like the human condition. But will that last? Probably not." Essentially, regardless of the banner they work under, they are working on a cure for aging.

Jan Vijg, who is the chair of the department of genetics at the Albert Einstein School of Medicine and a respected name in gerontology, wonders why his colleagues are so adamant that we can't cure aging. "It's very shortsighted," Vijg says. When we study the problem of aging and seek to slow it down, to prevent the suffering of old age, we are doing what we intended science and medicine to do from the beginning, and what we want science and medicine to do for us every day of our lives. How different, then, is the one goal from the other?

In Vijg's view, we may be able to extend the human life span significantly and soon. There are so many flaws in our bodies, thanks to our evolution as disposable soma, flaws we may be able to fix through SENS-like interventions. Babies born in 2030 may have a life expectancy of 120 and live halfway into the next century.

Vijg complains about the world's pessimistic view of aging every time I see him. Just imagine what would happen, he says, if the people at the National Cancer Institute were to announce, "We don't want to cure cancer, just make your last days more comfortable." They wouldn't get very far. But look, Vijg cries: that's just the way they talk at the National Institute of Aging!

In Vijg's opinion, Aubrey got broadsided by the gerontologists because he was the first to declare war on aging, cogently and forthrightly, in our present moment. That's why so many—particularly the older gerontologists—are furious with Aubrey, says Vijg. "He scooped them, in a sense."

Vijg wrote this to me the other day: "It's my impression that you have more difficulty than me in believing that we will really get there, and I am the scientist. I disagree with Aubrey on a lot of things, but I do think it's entirely rational to expect that we will be able to cure aging. While his ideas are often smart and he knows his stuff (I, like you, think he is quite brilliant), he lacks insight into the little things. And the little things in science are the reason I am now in 2009 about where I expected to be in 1989. He lacks insight into the hidden issues that often prevent projects from succeeding."

In Vijg's view, what will block our progress is the accumulation of mutations in our aging cells. In his recent book *Aging and the Genome*, Vijg argues that this load of mutations will be the wall we hit in the end. That will be "the ultimate limit to life," he writes. "Genomes cannot be cleansed of all genetic damage, because it is their nature to change and undergo mutation. Indeed, to keep genomes free of change would be to tamper with the logic of life itself."

Aubrey is not the only one who has tried to make an end run around this problem. The French novelist Michel Houellebecq imagines in his sardonic novel *The Elementary Particles* that a molecular biologist will soon discover a way to rewrite our genetic code into a form that does not mutate. The result will be a new species of clones who live forever in perfect tranquillity, without war, sex, identity, perversity, aging, or disease.

That scenario amuses Vijg. He concludes his book, "I think most of us would prefer to live a bit shorter, with our imperfect selves and an aging genome."

Almost a thousand years ago a scholar of Chinese poetry, Ssu-ma Kuang, praised a line by the poet Li Ho, who had died a few centuries before him, in the year 817, at age twenty-six. The scholar wrote, "Li Ho's *If heaven too had passions even heaven would grow old* is a peerless line." The scholar also noted that a poet of his own day, Shi Yen-nien, had paid homage to that wonderful line with another, "which people think is a close rival to it": *If the moon knew no yearning the moon would always be round.*

In the year 1760, a second scholar of Chinese poetry, Wang Ch'i, begged to disagree. "I have maturely considered the two lines; they exhibit the whole difference between the natural and the forced; there is no comparison between them."

The dews of Han are gone, along with the pans, the statues, the emperor, the dynasty, and the poets, but people still read those old lines and weigh them in the balance. That is as close to immortality as any mortal can come at present, as poets themselves eternally remind us. Théophile Gautier writes, "The Gods themselves die out, but Poetry, stronger even than bronze, survives everything."

In a way it is a common gift, to conquer time within our mortal life. Lovers know this. They feel it solemnly: at once mortal and immortal. Couples experience it even in their very dailiness, in the choice endlessly renewed, the pledge freely given. In the very sacrifice, eternity, as strange as that sounds. Here we all swim in the same ocean, mortalists and immortalists. "Who knows this, daily enjoys the Kingdom of Heaven," as it is written in the Upanishads.

Aubrey's faith that we are all of us, all mortals on spaceship Earth, in the middle of liftoff seems unshakable.

"Aubrey, *come on*!" I said to him the other day, when he was insisting to me once again that we can achieve escape velocity in our lifetime—that we are already blasting off.

"No '*come on*,'" he said sternly. He wanted to know what made me think we couldn't do it. What made me so sure we weren't doing it now? He stared back at me as fiercely as a bearded pirate, not a flicker of doubt in his eyes. He looked absolutely unmovable. He is getting older now himself; there are silver threads among the brown, when the strong sun falls on his beard. The first wrinkles, the first age spots. He's had a parting of the ways with the Methuselah Foundation; now he runs the SENS Foundation. Immortality is the cause to which he has given his youth.

When this is all over, he told me, and someone has produced a way to rejuvenate a mouse, he is retiring to Madagascar.

"Madagascar?"

"I've had quite enough of this, I can assure you."

I remember when I first came across Aubrey's extraordinary name. I read a little squib about him on the Web—one of his first pieces of publicity. I sent him an e-mail, he wrote back more or less instantly, and soon there he was in my office chair, beer bottle in hand, telling me that we could live forever.

I warned Aubrey that I did not find his ideas completely kosher. I said it that day and many times thereafter. On the road to Ravenna he looked so much like Jesus, loping along, that I reminded him I wasn't a disciple, that he and I were not of the same faith. I quoted one of my father's Yiddish proverbs. *"Az mih esst chazer, luzz rinnen ueber dem boord!"* If you're going to eat pork, let it dribble over your beard.

That's why I was talking with him and not with a conventional gerontologist, I said.

Aubrey laughed. "In for a penny, in for a pound," he said. "Jolly good."

Aubrey said I was crazy to doubt him—either crazy, or a lightweight. Sometimes at the Eagle the very sight of me seemed to exasperate him as I sat there smiling in my strained, coffee-stained sobriety. "It's just biz*arre*!" he would cry, like a lawyer at the bench, with a somewhat practiced indignation, his hands flying upward and his voice rising half an octave. "It's brainwashing! What other explanation *is* there?"

Meanwhile, from a stool at the far end of the bar, that old Cambridge codger kept staring at our table, the man with the fixed leer from the Mad Hatter's tea party, as if to say: You fools will never make sense. Whatever it is you are talking about, you will never make sense.

I still do not expect to see our last night's day, that dawn when immortality shall be unveiled, to the cry of the peacocks. I think our last talk in the Eagle was the moment when that was settled between us.

"I know a lot more about biology than you do," Aubrey declared, very stiffly, with the late afternoon sun casting those apocalyptic shafts of light onto his face. "So any conclusions based on that assumption are just illogical. You're preferring your own uninformed speculations as a nonbiologist to mine."

And yet, now that the journey is over, I'm surprised to find myself wavering. Maybe I'm half of the devil's party without knowing it.

One of my best memories of Aubrey is the brief walk we took after we had paid and left the pub, debating on the cobbles and flagstones. Aubrey showed me Trinity Hall, where he went to school, where he learned to punt, and where the ladder that led down to the river used to lean. He apologized for being so hard on me back at the Eagle. We agreed to disagree. He walked me partway to the train station through the English summer drizzle.

"Best of luck."

After we said our goodbyes, Aubrey sprinted off. He was late for dinner with Adelaide; he'd promised to be back at seven. He flew home down that old stone street like a schoolboy of twelve.

Acknowledgments

Aubrey de Grey was cheerful and extremely generous with his time and help, even though I told him from the outset that I was not writing as an acolyte. I'm very grateful.

Nick Lemann, dean of Columbia University's Graduate School of Journalism, gave me a crucial year's writing leave to finish the book. A fellowship from the John Simon Guggenheim Foundation helped make that year possible.

Eric Kandel, Arnie Levine, Paul Nurse, Martin Raff, Frank Rothman, and Harold Varmus gave me early advice and encouragement.

John Bonner, Stuart Firestein, Philip Kitcher, Nick Lemann, and Michael Shapiro read early drafts. Judy Campisi, Joan Finkelstein, Steve Helfand, Martin Raff, and Jan Vijg read late drafts. Many thanks to all of them for valuable suggestions and corrections.

Dozens of scientists and science watchers went out of their way to help, although not all of their names appear in the book. Special thanks to Martin Ackermann, Richard Cohen, Ralph Greenspan, Marguerite Holloway, and Neil Patterson.

Fay Schopen helped me with research, particularly in Cambridge. I still owe her a pint.

At Ecco, Dan Halpern saw the book's possibilities and waited for it with the patience of Job. When Matt Weiland joined Ecco a little over a year ago, he took on the day-by-day job of working with book and author. Matt shepherded these chapters out of many blind valleys. Without him, the book would not have found the path it has.

My agent, Kathy Robbins, was wonderful, as always. I feel lucky to work with her, and with the excellent staff of the Robbins Office.

My sons, Aaron and Benjamin, took an interest in this project from the first day, when Aubrey de Grey came to visit us. I owe them thanks for their good words and advice.

As always, my wife, Deborah Heiligman, read many drafts, taking time away from her own writing projects. She saw me through this book at some cost to her own longevity. In her, the meaning of the span.

Notes on Sources and Further Reading

"The subject is really an enormous subject," William James wrote in 1898, in his essay "Human Immortality." "At the back of Mr. Alger's *Critical History of the Doctrine of a Future Life* there is a bibliography of more than five thousand titles of books in which it is treated."

On top of that enormous subject, we now have the modern science of longevity. If you search for the keyword "gerontology" in the world's largest online index of medical literature, Medline, you get a list of more than 25,000 articles, all published since the year 1950.

Of course, longevity and immortality are not the same thing. Even if gerontologists learned to slow, stop, or even reverse the process of aging, they would not make human bodies live forever. They would eliminate only one cause of death. Nevertheless, aging is by far the most common cause of death in this vale of tears. A cure for aging would mean so many centuries or millennia of future life that the prospect looks very much like immortality from here.

Because the science of longevity is so young and turbulent, it's too soon for critical histories and giant bibliographies. Here are a few notes on some of my sources, chapter by chapter, with suggestions for further reading.

CHAPTER 1: IMMORTAL LONGINGS

In the last decade or so, as their field has heated up, gerontologists have published a whole shelf of books for a general audience. These include:

Austad, S. N. (1999). *Why We Age: What Science Is Discovering About the Body's Journey Through Life*. Wiley.

Butler, R. N. (2008). *The Longevity Revolution: The Benefits and Challenges of Living a Long Life*. PublicAffairs.

Carnes, B. A., and S. J. Olshansky (2002). *The Quest for Immortality: Science at the Frontiers of Aging*. Norton.

Guarente, L. (2002). *Ageless Quest: One Scientist's Search for the Genes That Prolong Youth*. Cold Spring Harbor Laboratory Press.

Hayflick, L. (1994). *How and Why We Age*. Ballantine.

Kirkwood, T. (1999). *Time of Our Lives: The Science of Human Aging*. Oxford University Press.

Rose, M. R. (2005). *The Long Tomorrow: How Advances in Evolutionary Biology Can Help Us Postpone Aging*. Oxford University Press.

West, M. (2003). *The Immortal Cell: One Scientist's Quest to Solve the Mystery of Human Aging*. Doubleday.

Aubrey de Grey has published a book about his "Strategies for Negligible Senescence":

de Grey, A. D. N. J., with Michael Rae (2007). *Ending Aging: The Rejuvenation Breakthroughs That Could Reverse Human Aging in Our Lifetime*. St. Martin's.

Aubrey de Grey has also published almost one hundred manifestos and scientific articles on the subject. For one of the early, provocative papers in which he rode out to battle against most gerontologists, see de Grey, A. D., B. N. Ames, et al. (2002). "Time to talk SENS: Critiquing the immutability of human aging." *Ann N Y Acad Sci* 959: 452–62; discussion 463–65. As de Grey and his coauthors write, "Aging is a three-stage process: metabolism, damage, and pathology. The biochemical processes that sustain life generate toxins as an intrinsic side effect. These toxins cause damage, of which a small proportion cannot be removed by any endogenous repair process and thus accumulates." Finding ways to remove the accumulating damage, they argue, "would sever the link between

chapter. If you dig down into any decade, you can find a dozen now forgotten doctors and biologists who hoped to live forever. A few months ago I found an old, slightly pulpy, but entertaining paperback: McGrady, P. M., Jr. (1968). *The Youth Doctors.* Ace. It is full of names of lost immortalists, including one rebel whose polemics have at least a family resemblance to Aubrey de Grey's.

See also Comfort, A. *The Process of Ageing* (1964). Signet Science Library. Dated, but still good reading.

Another readable paperback from that time, also with yellowing pages:

Harrington, A. (1969). *The Immortalist: An Approach to the Engineering of Man's Divinity.* Avon. It begins, "Death is an imposition on the human race, and no longer acceptable."

CHAPTER 3: LIFE AND DEATH OF A CELL

Here are a few books about the beauty of the beginning of the life cycle:

Bonner, J. T. (1993). *Life Cycles: Reflections of an Evolutionary Biologist.* Princeton University Press. John Tyler Bonner, born in 1920 and still going strong, is one of the best biologist-writers alive. This is a delightful book about the evolution of the life cycle, and the evolution of Bonner.

Gilbert, S. F. (2006). *Developmental Biology.* Sinauer. The standard textbook.

Wolpert, L. (1991). *The Triumph of the Embryo.* Oxford University Press.

On the evolution of multicellular life:

Bonner, J. T. (2000). *First Signals: The Evolution of Multicellular Development.* Princeton University Press.

Buss, L. W. (1987). *The Evolution of Individuality.* Princeton University Press. A bit old, and difficult, but fascinating.

This monumental work of scholarship helped bring new life to the science of gerontology:

Finch, C. E. (1990). *Longevity, Senescence, and the Genome*. University of Chicago Press.

On the hydra:

Martinez, D. E. (1998). "Mortality patterns suggest lack of senescence in hydra." *Exp Gerontol* 33: 217–25.

Maria Rudzinska was working in a long tradition at Rockefeller University. Not only was Alex Carrel there before her, so was another early eminence there, Jacques Loeb. See, for instance:

Loeb, J., and J. H. Northrup (1917). "On the influence of food and temperature upon the duration of life." *Biological Chem*. 32: 103–21.

Two of Rudzinska's papers on her beloved *Tokophrya*:

Rudzinska, M. A. (1951). "The influence of amount of food on the reproduction rate and longevity of a suctorian (*Tokophrya infusionum*)." *Science* 113: 10–11.

Rudzinska, M. A. (1984). "Cellular and clonal aging in the suctorian protozoan Tokophrya infusionum." S. J. Karakashian, H. N. Lanners, and M. A. Rudzinska. *Mech. Ageing Develop*. 26: 217–29.

CHAPTER 4: INTO THE NEST OF THE PHOENIX

A wonderful and authoritative collection of old Jewish legends:

The Book of Legends, Sefer Ha-Aggadah: Legends from the Talmud and Midrash (1999). H. N. Bialik, editor; W. Braude, translator. Schocken.

For reviews of Denham Harman's thinking about aging, see:

Harman, D. (2006). "Free radical theory of aging: an update," *Ann NY Acad Sci* 1067: 10–21.

Kitani, K., and G. O. Ivy (2003). "'I thought, thought, thought for four months in vain and suddenly the idea came.' Interview with Denham and Helen Harman." *Biogerontology* 4: 401–12.

de Grey, A. D. (1997). "A proposed refinement of the mitochondrial free radical theory of aging." *Bioessays* 19(2): 161–66.

de Grey, A. D. (2002). "Three detailed hypotheses implicating oxidative damage to mitochondria as a major driving force in homeotherm aging." *Eur J Biochem* 269(8): 1995.

CHAPTER 5: THE EVOLUTION OF AGING

Peter Medawar's key essays about the evolution of aging are reprinted in:

Medawar, P. (1981). *The Uniqueness of the Individual*. Dover.

He returns to the subject in his quirky autobiography:

Medawar, P. (1988). *Memoir of a Thinking Radish*. Oxford University Press.

And again in:

Medawar, P. (1990). *The Threat and the Glory*. HarperCollins.

See also:

Finch, C. E., and E. M. Crimmins (2004). "Inflammatory exposure and historical changes in human life-spans." *Science* 305(5691): 1736–39.

Caspari, R., and Lee, S.-H. (2004). "Older age becomes common late in human evolution." *PNAS* 101(30): 10895–10900.

Crespi, B. J. (2004). "Vicious circles: Positive feedback in major evolutionary and ecological transitions." *TREE* 19(12): 627–33.

Rose, M. R. (1991). *Evolutionary Biology of Aging*. Oxford University Press.

Stearns, S. C., and J. C. Koella (2008). *Evolution in Health and Disease*. Oxford University Press.

Platt, R. (1963). "Reflections on ageing and death." *Lancet*. 281: 1–6.

CHAPTER 6: THE GARBAGE CATASTROPHE

Holliday, R. (2006). "Aging is no longer an unsolved problem in biology." *Ann NY Acad Sci* 1067: 1–9.

de Grey, A. D. (2007). "Alzheimer's, atherosclerosis, and aggregates: A role for bacterial degradation." *Nutr Rev* 65(12 Pt 2): S221–27.

Terman, A., and U. T. Brunk (2006). "Oxidative stress, accumulation of biological 'garbage,' and aging." *Antioxid Redox Signal* 8(1–2): 197–204.

Stroikin, Y., H. Dalen, et al. (2005). "Testing the 'garbage' accumulation theory of ageing: mitotic activity protects cells from death induced by inhibition of autophagy." *Biogerontology* 6(1): 39–47.

CHAPTER 7: THE SEVEN DEADLY THINGS

For a full-length treatment of de Grey's Strategies for the Engineering of Negligible Senescence, see his book *Ending Aging* (op cit.).

He's also published dozens of shorter accounts, including:

de Grey, A. D. (2005). "A strategy for postponing aging indefinitely." *Stud Health Technol Inform* 118: 209–19.

For Michael Hecht's experiments with beta-amyloid:

Kim, W., Y. Kim, et al. (2006). "A high-throughput screen for compounds that inhibit aggregation of the Alzheimer's peptide." *ACS Chem Biol* 1(7): 461–69.

Kim, W., and M. H. Hecht (2008). "Mutations enhance the aggregation propensity of the Alzheimer's A beta peptide." *J Mol Biol* 377(2): 565–74.

de Grey, A. D., P. J. Alvarez, et al. (2005). "Medical bioremediation: prospects for the application of microbial catabolic diversity to aging and several major age-related diseases." *Ageing Res Rev* 4(3): 315–38.

CHAPTER 8: THE METHUSELAH WARS

For a meticulously detailed contemporary history, see Hall, S. (2003). *Merchants of Immortality: Chasing the Dream of Human Life Extension.* Houghton Mifflin.

Klass, M. R. (1983). "A method for the isolation of longevity mutants in the nematode Caenorhabditis elegans and initial results." *Mech Ageing Dev* 22(3–4): 279–86.

Klass, M., and D. Hirsh (1976). "Non-ageing developmental variant of Caenorhabditis elegans." *Nature* 260(5551): 523–25.

Kenyon, C., J. Chang, et al. (1993). "A C. elegans mutant that lives twice as long as wild type." *Nature* 366(6454): 461–64.

Kenyon, C. (2005). "The Plasticity of Aging: Insights from Long-Lived Mutants." *Cell* 120: 449–60.

Song, S., and T. Finkel (2007). "GAPDH and the search for alternative energy." *Nature Cell Biology* 9(8): 869–70.

Zhang C., and A. M. Cuervo (2008). "Restoration of chaperone-mediated autophagy in aging liver improves cellular maintenance and hepatic function." *Nat Med* 14(9): 959–65.

Kaushik S., and A. M. Cuervo (2008). "Chaperone-mediated autophagy." *Methods Mol Biol* 445:227–44.

Cuervo, A. M. (2008). "Autophagy and aging: Keeping that old broom working." *Trends Genet* 24: 604–12.

Cuervo, A. M., L. Stefanis, et al. (2004). "Impaired degradation of mutant alpha-synuclein by chaperone-mediated autophagy." *Science* 305: 1292–95.

Mizushima N., B. Levine, et al. (2008). "Autophagy fights disease through cellular self-digestion." *Nature* 451:1069–75.

Hansen, M., A. Chandra, et al. (2008). "A role for autophagy in the extension of lifespan by dietary restriction in C. elegans." *PLoS Genet* 4(2): e24.

Rubinsztein, D. C., J. E. Gestwicki, et al. (2007). "Potential therapeutic applications of autophagy." *Nat Rev Drug Discov* 6(4): 304–12.

Sarkar, S., E. O. Perlstein, et al. (2007). "Small molecules enhance autophagy and reduce toxicity in Huntington's disease models." *Nat Chem Biol* 3(6): 331–38.

Sarkar, S., B. Ravikumar, et al. (2009). "Autophagic clearance of aggregate-prone proteins associated with neurodegeneration." *Methods Enzymol* 453: 83–110.

CHAPTER 9: THE WEAKEST LINK

Aubrey de Grey introduced his cancer cure in these papers:

de Grey, A. D. (2005). "Whole-body interdiction of lengthening of telomeres: a proposal for cancer prevention." *Front Biosci* 10: 2420–29.

de Grey, A. D., F. C. Campbell, et al. (2004). "Total deletion of in vivo telomere elongation capacity: an ambitious but possibly ultimate cure for all age-related human cancers." *Ann N Y Acad Sci* 1019: 147–70.

De Grey writes about his hopes for the Singularity in Edge.com, January 2, 2009. http://ieet.org/index.php/IEEET/more/2781.

See also de Grey, A. D. (2009). "The singularity and the Methuselarity: Similarities and differences." *Stud Health Technol Inform* 149: 195–202. Here he writes, "Aging, being a composite of innumerable types of molecular and cellular decay, will be defeated incrementally. I have for some time predicted that this succession of advances will feature a threshold, which I here christen the 'Methuselarity. . . .'" As he says, that threshold is very close to the Singularity.

For a massive defense of the Singularity idea, see Kurzweil, R. (2005). *The Singularity Is Near: When Humans Transcend Biology.* Viking.

CHAPTER 10: LONG FOR THIS WORLD

For an excellent collection of papers on aspects of the great questions "can we" and "should we," see:

Post, S. G., and R. H. Binstock, eds. (2004). *The Fountain of Youth: Cultural, Scientific, and Ethical Perspectives on a Biomedical Goal.* Oxford University Press.

For John Cheever's envy of Saul Bellow's immortality, see Atlas, J. (2000). *Bellow: A Biography.* Random House.

Updike's last book of poems is one of his finest: *Endpoint and Other Poems* (2009). Alfred A. Knopf.

Because life expectancy has lengthened within the lifetimes of baby boomers, they may find the approach of old age even more disturbing, in some ways, than generations past. What if their ship is sinking within sight of land? Shakespeare speaks of the "double death": "'Tis double death to drown in ken of shore."

For an insightful essay on baby boomers' competition for more years, see Kinsley, M. (2008). "Mine is longer than yours." *New Yorker* (April 7).

On the psychology of time—our own private expectations of how much or little time lies ahead—see Carstensen, L. L. (2006). "The influence of a sense of time on human development." *Science* 312(5782): 1913–15.

There is a large and growing literature on twenty-first-century demography. In this chapter, I cite Christensen, K., G. Doblhammer, et al. (2009). "Ageing populations: the challenges ahead." *Lancet* 374(9696): 1196–208. Christensen, K., A. M. Herskind, et al. (2006). "Why Danes are smug: comparative study of life satisfaction in the European Union." *Bmj* 333(7582): 1289–91.

For Aubrey de Grey's rejection of demographers' forecasts:

de Grey, A. D. (2006). "Extrapolaholics anonymous: Why demographers' rejections of a huge rise in cohort life expectancy in this century are overconfident." *Ann NY Acad Sci* 1067: 83–93.

Terman, A., and U. T. Brunk (2005). "Is aging the price for memory?" *Biogerontology* 6:205–10.

CHAPTER 11: THE TROUBLE WITH IMMORTALITY

For the story of Luz, I consulted *The Book of Legends*, Bialik and Braude (op cit.).

This book is still worth reading, although it is dated by its conviction that Freud and his disciples had us figured out: Becker, E. (1973). *The Denial of Death.* Free Press.

For a wonderful new book exploring some of the same emotional territory, read Barnes, J. (2008). *Nothing to Be Frightened Of.* Alfred A. Knopf.

Veatch, R. M. (2009). "The evolution of death and dying controversies." *Hastings Center Report* 39(3): 16–19.

Haeckel, E. (1900). *The Riddle of the Universe.* Harper & Brothers.

Čapek, K. (1925). *The Makropulos Secret.* International Pocket Library.

Bernard Williams's paper about Čapek's play is reprinted here:

Williams, B. (1973). *Problems of the Self.* Cambridge University Press.

Someday soon, neurophilosophers may be able to explore Joshua Lederberg's point about remembering and forgetting. See, for instance:

Shuai, Y., B. Lu, et al. (2010). "Forgetting is regulated through Rac activity in Drosophila." *Cell* 140(4): 579–89.

And see Bhanoo, S. N. (2010). "Forgetting, with a purpose." *New York Times.*

If we ever do stop aging, what will become of the wisdom of the ages? Everything ever written about childhood, youth, middle and old age will seem incredibly dated. For the time being, at least, this is still a valuable anthology:

Sampson, A., and S. Sampson (1985). *The Oxford Book of Ages.* Oxford University Press.

CHAPTER 12: THE EVERLASTING YES AND NO

Robert Butler writes about the "longevity dividend" in his book *The Longevity Revolution* (op cit.). See also Butler, R. N., R. A. Miller, et al. (2008). "New model of health promotion and disease prevention for the 21st century." *BMJ* 337: a399.

Two attacks from the center against the fringes of antiaging medicine:

Olshansky, S. J., L. Hayflick, et al. (2002). "Position statement on human aging." *J Gerontol A Biol Sci Med Sci* 57(8): B292–97.

Olshansky, S. J., L. Hayflick, et al. (2002). "No truth to the fountain of youth." *Sci Am* 286(6): 92–95.

Vijg, J. (2007). *Aging of the Genome: The Dual Role of DNA in Life and Death.* Oxford University Press.

Houellebecq, M. (2001). *The Elementary Particles*. New York, Vintage.

Graham, A. C., trans. (2008). *Poems of the Late T'ang*. NYRB Classics.

William Butler Yeats composed beautiful translations of the Upanishads. Yeats, W. B., and Swami Shree, trans. (1975). *The Ten Principal Upanishads*. Macmillan.

Index